Azure Strategy and Implementation Guide, Fourth Edition

The essential handbook to cloud transformation with Azure

Jack Lee

Greg Leonardo

Jason Milgram

Dave Rendón

BIRMINGHAM—MUMBAI

Azure Strategy and Implementation Guide, Fourth Edition

Authors: Jack Lee, Greg Leonardo, Jason Milgram, and Dave Rendón

Technical Reviewers: Aaditya Pokkunuri and Peter De Tender

Managing Editors: Siddhant Jain and Mamta Yadav

Acquisitions Editors: Saby Dsilva and Ben Renow-Clarke

Production Editor: Deepak Chavan

Editorial Board: Vishal Bodwani, Ben Renow-Clarke, Arijit Sarkar, and Lucy Wan

Third Published: June 2020

Fourth Published: May 2021

Production Reference: 4050521

ISBN: 978-1-80107-797-2

Published by Packt Publishing Ltd.

Livery Place, 35 Livery Street

Birmingham, B3 2PB, UK.

Table of Contents

Chapter 3: Modernizing with hybrid cloud and multicloud   31

Chapter 4: Cloud migration: Planning, implementation, and best practices 75

Chapter 6: Security fundamentals to help protect against cybercrime

Preface

About

This section briefly introduces the authors and reviewers, the coverage of this book, the technical skills you'll need to get started, and the hardware and software required to complete all of the topics.

About Azure Strategy and Implementation Guide, Fourth Edition

Microsoft Azure is a powerful cloud computing platform that offers a multitude of services and capabilities for organizations of any size pursuing a cloud strategy.

This fourth edition discusses the latest updates on security fundamentals, hybrid cloud, cloud migration, Microsoft Azure Active Directory, and Windows Virtual Desktop. It encapsulates the entire spectrum of measures involved in Azure deployment, including understanding Azure fundamentals, choosing a suitable cloud architecture, building on design principles, becoming familiar with Azure DevOps, and learning best practices for optimization and management.

The book begins by introducing you to the Azure cloud platform and demonstrating the substantial scope of digital transformation and innovation that can be achieved with Azure's capabilities. It then provides practical insights on application modernization, Azure **Infrastructure as a Service (IaaS)** deployment, infrastructure management, key application architectures, best practices of Azure DevOps, and Azure automation.

By the end of the book, you will have acquired the essential skills to drive Azure operations from the planning and cloud migration stage to cost management and troubleshooting.

About the authors

Jack Lee is a senior Azure certified consultant and an Azure practice lead with a passion for software development, cloud, and DevOps innovations. He is an active Microsoft tech community contributor and has presented at various user groups and conferences, including the Global Azure Bootcamp at Microsoft Canada. Jack is an experienced mentor and judge at hackathons and is also the president of a user group that focuses on Azure, DevOps, and software development. He is the co-author of *Azure for Architects and Cloud Analytics with Microsoft Azure*, published by Packt Publishing. He has been recognized as a Microsoft MVP for his contributions to the tech community. You can follow Jack on Twitter at @jlee_consulting.

Greg Leonardo is currently a cloud architect, helping organizations with cloud adoption and innovation. He has been working in the IT industry since his time in the military. He is a veteran, architect, teacher, speaker, and early adopter. Currently, he is a Certified Azure Solution Architect Expert, Microsoft Certified Trainer, Microsoft Azure MVP, and has worked in many facets of IT throughout his career. He is President of TampaDev, a community meetup that runs #TampaCC, Azure User Group, Azure Medics, and various technology events throughout Tampa.

Greg has also authored the book *Hands-On Cloud Solutions with Azure* and the previous two editions of *Azure Strategy and Implementation Guide for Microsoft* by Packt Publishing.

Jason Milgram is a chief architect at **Science Applications International Corporation (SAIC)** headquartered in Reston, VA. Previously he was the first VP Cloud Solution Architect at City National Bank of Florida in Miami, and before that he was VP of Platform Architecture and Engineering at Champion Solutions Group in Boca Raton, Florida. Jason was educated at the University of Cincinnati and the Massachusetts Institute of Technology – Sloan School of Management. He was also a Sergeant in the US Army Reserve, serving from 1990 to 1998. A Microsoft Azure MVP (2010-present), Jason has given over 100 Azure presentations and regularly writes articles on Azure topics.

Dave Rendón has been a Microsoft MVP for 7 consecutive years with expertise in Azure, currently working as a solutions architect for the US and LATAM regions at Kemp, a provider of application delivery software and security.

He regularly presents at public IT events such as Microsoft Ignite, Global Azure events, and local user group events across the US, Europe, and Latin America, and is active on Twitter as @daverndn.

Dave has had a strong focus on Microsoft technologies and Azure since 2010. He helps people develop in-demand skills to advance their career through the cloud and AI and provides support to Microsoft partners worldwide on technical guidance and provides Azure training classes globally (India, South America, and the US) that help companies migrate critical applications to the cloud and train their staff to be certified cloud architects.

You can reach Dave as /daverndn on LinkedIn and Twitter, and as /wikiazure on YouTube.

About the reviewers

Aaditya Pokkunuri is an experienced senior database engineer with a demonstrated history of working in the information technology and service industry, with over 11 years of experience. He is skilled in performance tuning, Microsoft SQL Database Server Administration, SSIS, SSRS, PowerBI, and SQL development. He possesses strong knowledge of replication, clustering, SQL Server high availability options, and ITIL processes. His expertise lies in Windows administration tasks, Active Directory, and Microsoft Azure technologies. He has expertise in AWS Cloud and is an AWS Solution Architect Associate as well. Aaditya is a strong information technology professional with a Bachelor of Technology degree focused on computer science and engineering from Sastra University, Tamil Nadu.

Peter De Tender is a well-known Azure expert and a very passionate and dedicated technical trainer, who always manages to provide inspiring, deeply technical workshops on the Azure platform, packed with demos and fun.

Before Peter joined the prestigious Azure Technical Trainer team within Microsoft, he held a similar position in his own company for 6 years. Now, he's continuing what he loves doing most, upskilling customers and partners in the wonderful world and capabilities of Azure.

Peter has been a **Microsoft Certified Trainer** (**MCT**) for more than 10 years and has been a Microsoft MVP since 2013, initially on Windows IT Pro, but he moved to the Azure category in 2015.

Besides co-authoring the previous edition of this book, Peter has published other Azure-oriented material with Packt Publishing, Apress, and through self-publishing.

You can follow Peter on Twitter as @pdtit or @007ffflearning to stay up to date with his Azure adventures.

Learning objectives

By the end of this cookbook, you will be able to:

- Understand core Azure infrastructure technologies and solutions
- Carry out detailed planning for migrating applications to the cloud with Azure
- Deploy and run Azure infrastructure services
- Define roles and responsibilities in DevOps
- Get a firm grip on security fundamentals
- Carry out cost optimization in Azure

Audience

This book is designed to benefit Azure architects, cloud solution architects, Azure developers, Azure administrators, and anyone who wants to develop expertise in operating and administering the Azure cloud. Basic familiarity with operating systems and databases will help you grasp the concepts covered in this book.

Approach

The *Azure Strategy and Implementation Guide, Fourth Edition* explains each topic in detail and provides a real-world scenario at the end to acquaint you with practical solutions. It also adds value to the lessons you learn with supplemental statistical data and graphical representations.

Hardware requirements

The Azure portal is a web-based console and runs on all modern browsers for desktops, tablets, and mobile devices. To use the Azure portal, you must have JavaScript enabled on your browser.

Software requirements

We recommend that you use the most up-to-date browser that's compatible with your operating system. The following browsers are supported:

- Microsoft Edge (latest version)
- Internet Explorer 11
- Safari (latest version, Mac only)
- Chrome (latest version)
- Firefox (latest version)

Conventions

Code words in the text, database names, folder names, filenames, and file extensions are shown as follows: "In this code, we are using a `let` statement for the `Event` called `WVDConnections`, and filtering tables for rows that match the users with `connected` state."

Here's a sample block of code:

```
let Events = WVDConnections
    | where UserName == "userupn";
Events
| where State == "Connected"
| project CorrelationId, UserName, ResourceAlias, StartTime=TimeGenerated
| join (Events
    | where State == "Completed"
    | project EndTime=TimeGenerated, CorrelationId)
    on CorrelationId
| project Duration = EndTime - StartTime, ResourceAlias
| sort by Duration asc
```

New terms and important words are shown in **bold**. For example, "Edge computing combines the power of the cloud with **Internet of Things (IoT)** devices."

1
Introduction

Today's businesses face significant challenges, from enabling remote work and responding to increased cyberattacks to managing reduced cash flow. Microsoft Azure is a powerful cloud platform designed to help you empower productivity, ensure security, drive efficiency, and save money, enabling you to deliver resiliency, cost savings, and the impact your company requires. Whether you are a start-up or a multinational enterprise operating all over the world, you can start deploying and migrating workloads to Azure with an approach that meets your business needs.

The first step in taking advantage of the many capabilities offered by Azure is careful planning. This book has been created to help you undertake this planning successfully by providing you with a foundational understanding of Azure infrastructure, its core capabilities and benefits, and best practices that will help you successfully use Azure, whether you decide to fully migrate or run a hybrid cloud approach.

This chapter will cover the following:

- What is Microsoft Azure?
- Approaches for Azure adoption
- Azure migration strategies
- Business benefits of Azure infrastructure

This chapter will introduce you to the framework and an overview of the business benefits of Azure. The following chapters will build on that, providing more specific guidance and an explanation of the technologies that will help you plan your migration and Azure infrastructure implementation strategies.

What is Microsoft Azure?

Azure is Microsoft's cloud computing platform. It provides a variety of services to both individuals and organizations. Cloud computing enables convenient, on-demand access to a shared pool of computing resources over the network. These resources can range from storage and servers to applications that can be deployed rapidly.

Azure provides rapid provisioning of compute resources to help you host your existing applications, streamline new application development, and even enhance on-premises applications. These resources are managed by Microsoft; however, you can monitor them and get reports and alerts when issues arise. This is all built on top of resource pools that can be dynamically assigned to the required services, which can include CPU, memory, storage, and network bandwidth.

There are four categories of service models offered by Azure, which are **Infrastructure as a Service (IaaS)**, **Software as a Service (SaaS)**, **Platform as a Service (PaaS)**, and Serverless. Azure IaaS is an instant computing infrastructure that offers essential compute, storage, and networking resources on demand and is provisioned and managed over the internet.

It is critical to understand how you can leverage each of these service models to meet your ever-changing demands. When you use Azure, you have a shared responsibility with regard to the resources you deploy. In *Figure 1.1*, you can see the extent to which you share and manage workload responsibilities with Microsoft for each of the service models, allowing you to focus on resources that are important for your application:

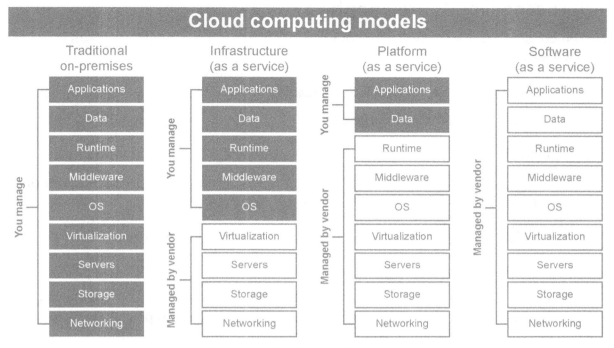

Figure 1.1: Cloud computing models

Depending on the level of responsibilities you'd like to manage compared to what Azure manages, you can determine which cloud service model works best for your organization. Different models afford you different degrees of control over the development environment, the ability to refactor your apps, your time-to-market goals, and so on. Your responsibilities generally increase as you move from SaaS to PaaS to IaaS. The separation of responsibilities will be covered in depth in *Chapter 2, Automation and governance in Azure*, in the *Identity and access control* section.

For the remainder of this book, we will be focusing on IaaS. IaaS gives you the most control over the environment as Microsoft provides the foundational infrastructure while you manage the applications. This approach is great for organizations that are looking to customize their cloud solutions for their business applications.

Approaches for Azure adoption

As you consider Azure adoption for your cloud infrastructure strategy, you can choose from different ways to deploy cloud services—public cloud, private cloud, hybrid cloud, multiple clouds, and at the edge of networks. Deciding between these computing environments can depend on a variety of factors.

Let's take a look at each of these approaches.

Public cloud

The most common type of cloud computing service is a public cloud, which is provided by third-party cloud providers and delivered over the public internet. The resources in these public cloud environments are owned and managed by the underlying cloud provider, which in the case of Azure is Microsoft. In public clouds, resources are shared with other organizations or cloud tenants; these can include services such as email, CRM, VMs, and databases.

Some common public cloud use cases involve organizations who want to expose their public workloads, such as public websites or mobile apps. These types of applications have workloads with multiple layers of UI and services in order for them to function. A good example of this is Microsoft's Office products on different platforms, such as desktop, web, and mobile. These offerings are accomplished with the UI being separate from the services, with each implementation using the services and cloud infrastructure behind the scenes to deliver the same experience regardless of the platform.

Azure offers several advantages as a public cloud:

- Azure is built on a vast network of servers and regions, which helps to protect against failures and guarantees high reliability.

- Azure can achieve near-unlimited scalability by providing on-demand services to meet your organization's needs.

- Azure provides both hardware and software at much lower costs compared to on-premises models, as you pay on a consumption basis.

Private cloud

As the name suggests, a private cloud comprises cloud computing resources—hardware and software—used exclusively by one business or organization, with both services and infrastructure being maintained on a private network. A third-party service provider can host a private cloud, or it can be located at an on-site datacenter.

Private clouds are often used by government agencies, financial institutions, or healthcare providers to meet specific regulatory and IT requirements with business-critical operations seeking enhanced control over their environment. On-premises environments can expand to Azure by using Azure ExpressRoute or a site-to-site VPN tunnel to connect workloads while keeping them isolated from public view. We will take a deeper look into these core Azure infrastructure components in *Chapter 5, Enabling secure, remote work with Microsoft Azure AD and WVD*.

The advantages of choosing a private cloud are:

- Greater control over your resources as they are not shared with others.
- Greater flexibility in customizing the environment to meet specific business needs.
- Better scalability when compared to on-premises infrastructures.

These types of cloud-based solutions are a bit more costly because of the infrastructure needed to isolate and access the organization's workloads.

Hybrid cloud

A hybrid cloud is a computing environment that combines an on-premises datacenter with a public cloud, allowing data and applications to be shared between them. Hybrid clouds allow businesses to seamlessly scale up their on-premises infrastructure into the cloud when processing demand increases and scale back that infrastructure when demand decreases. Hybrid cloud also allows the flexibility to use new cloud-first technologies for new or migrated workloads while keeping other business-critical applications and data on-premises due to migration costs and business or regulatory compliance.

To illustrate, let's think about a tax company scenario. Tax companies generally need large compute only three months out of the year, which can be costly. Rather than investing a large amount of capital in additional on-premises servers to support peak capacity, they can use a hybrid environment in the cloud to expand and contract their compute on demand. This would allow them to keep costs down by only paying for what they use.

Azure offers unique hybrid capabilities that give customers the flexibility to innovate anywhere, whether on-premises, across clouds, or in edge environments. There are different hybrid usage models in Azure that can help you reduce the cost of running your workloads in the cloud. For example, Azure Hybrid Benefit lets you bring your existing on-premises server licenses to Azure to maximize cost savings. This is referred to as the Hybrid licensing model and applies to most server-based licensing.

Microsoft also has unique industry-specific cloud offerings, such as Microsoft Cloud for Financial Services, Microsoft Cloud for Manufacturing, and Microsoft Cloud for Nonprofit. These industry clouds marry the whole range of Microsoft's cloud services with industry-specific components and standards, workflows, and APIs to provide tailored solutions to industry challenges, so businesses can target the areas that require technological transformation the most.

Hybrid clouds should not be considered a temporary middle ground that organizations inhabit only while they transition from on-premises to Azure. Instead, a hybrid cloud can be a strategy employed by organizations to find a stable solution that best addresses their information-technology needs.

Multicloud

A multicloud approach involves the use of multiple cloud computing services from more than one cloud provider. This allows you to mix and match services from different providers to get the best combination for a particular task or capitalize on offerings in specific locations, whether they are public or private clouds.

For instance, customers may choose a multicloud strategy to meet regulatory or data sovereignty requirements in different countries. This may also be done to improve business continuity and disaster recovery, for example, backing up on-premises data in two public clouds for business units, subsidiaries, or acquired companies that adopt different cloud platforms.

Multicloud models can become very complex as they require management across multiple platforms. Microsoft Azure provides solutions to help you operate your hybrid cloud seamlessly, which you can read about at https://azure.microsoft.com/solutions/hybrid-cloud-app/. One such solution is Azure Arc, a multicloud management technology. Azure Arc extends Azure management and services with a single control plane across hybrid, multicloud, and edge environments, enabling a consistent state across resource environments and infrastructure. It provides greater visibility of resources, team accountability, and developer empowerment while accelerating innovation from Azure to any location.

Both hybrid cloud and multicloud solutions will be covered in more detail in *Chapter 3, Modernizing with hybrid cloud and multicloud.*

Edge computing

Edge computing combines the power of the cloud with **Internet of Things (IoT)** devices. At edge locations, close to where the data resides, you can run virtual machines, data services, and containers using edge computing to gain insights in real time and reduce latency. At the edge of the network, your devices spend less time communicating with the cloud and operate reliably even through extended offline periods.

The wide adoption of smart sensors and connected devices, along with state-of-the-art cloud technologies such as machine learning and AI, means IoT devices are highly responsive to local changes and are contextually aware. There are also security benefits given the distributed nature of edge computing systems, which makes it difficult for a single disruption to compromise the entire network.

This can become beneficial for something like fleet tracking. With the help of Azure, the **United Nations Development Programme (UNDP)** devised a fleet management solution using IoT technology and enabled devices to gain new insights. These IoT tracking devices send a significant amount of telemetry data when they are connected to the internet and store this data locally in the absence of a connection. This has allowed the UNDP to move and manage its fleet of vehicles as it coordinates around 12,000 staff members in a mission to eradicate poverty through sustainable development. To find out more, visit https://customers.microsoft.com/story/822486-united-nations-development-programme-nonprofit-azure-iot.

We have seen the different ways you can adopt Azure in your cloud infrastructure strategy, so let's discuss how Azure migration actually works in more detail.

Azure migration strategies

Microsoft Azure gives organizations the ability to push an already existing infrastructure to the cloud, moving either some of their workloads for a hybrid cloud approach, or the entire infrastructure—this is referred to as *migration*. From migrating legacy applications to deploying applications on Azure, organizations need to determine their requirements beforehand and plan a migration strategy.

Migrating to Azure can be accomplished in several ways based on two important considerations. The first is which type of deployment model you'd like to use: public Azure, private cloud, hybrid Azure, or multicloud. The second is the service category or type: IaaS, PaaS, SaaS, or Serverless. These migration strategies will help you understand which approach may be optimal for moving your workloads to Azure.

There are three different strategies to accomplish migration to Azure: re-hosting, re-platforming, and refactoring.

Re-hosting

Re-hosting, or lifting and shifting, is the process of taking an on-premises application host or VM and moving it directly to Azure. It is the fastest and easiest way to move because it has the fewest dependencies, the lowest business impact, and no constraints. Re-hosting your application is recommended in scenarios where speed of delivery is needed.

Re-platforming

Re-platforming, or redeploying, is when you want to take something like the **Internet Information Services** (**IIS**) on a VM and move it out to a PaaS offering in Azure. This means you won't need to manage the OS, just the application itself. It can generally be performed by simply redirecting your DevOps process to redeploy to the new infrastructure, though there can be third-party DLLs and restrictions that may cause problems in this scenario.

Refactoring

Refactoring is generally recommended when your application code isn't compliant with Azure services. It requires you to rewrite the application, or parts of the application, to conform to the ever-evolving new standards, as well as functional and security needs. It is also referred to as *modernization* of the app because you are making it more Azure native. Of the three strategies for moving to Azure, this one has the potential to have cost-overrun risk.

You should note that there is no right or wrong answer when it comes to choosing between these approaches. Identifying your business goals can help you determine which migration strategy works best for your organization.

Let's talk about why you might consider performing a migration to Azure.

Business benefits of Azure infrastructure

When considering your adoption of Azure infrastructure, it's helpful to understand the benefits you can realize as you make the case for your organization to migrate. The flexible and agile nature of Azure is something that an on-premises infrastructure environment simply cannot match. Furthermore, by migrating to Azure, you gain the following benefits:

- **Scalability**: Azure can take care of the operational work and allow you to quickly scale up and down Azure resources to meet your business demands. You can easily provision new resources and scale up and down existing resources through the Azure portal, or programmatically via Azure PowerShell, the **Azure command-line interface** (**Azure CLI**), or REST APIs.

- **Cost savings**: Azure offers cloud services on a pay-as-you-go basis, allowing organizations to move from a CapEx to a more agile OpEx spending model. Save money and achieve operational agility with hybrid offers, comprehensive datacenter migration programs, and cost-optimized IT infrastructure.

- **Increased delivery speed**: Since you don't need to wait for infrastructure to be deployed in your datacenter to access the resources you need, you'll see an accelerated time to market. The automation of the continuous integration, delivery, and deployment pipeline on the single platform that is Azure DevOps also plays a big role in this.

- **Innovation**: You get access to all the latest technologies on Azure, such as Azure AI, Machine Learning, and IoT.

- **Seamlessly and securely manage hybrid environments**: You can start taking advantage of Azure-based resources without having to fully migrate your entire existing on-premises infrastructure to the cloud. Furthermore, you can apply security and resiliency across these hybrid environments. Azure allows you to improve your security posture and get comprehensive insights into the threats facing your environments.

As mentioned, Azure can provide resources on-demand and operates under a consumption-based model, which means you only pay for what you use. You need no longer invest in on-premises resources and applications; Azure allows you to infinitely scale your applications to perform better, with security at its forefront, benefiting small and large organizations alike.

Now that we have an overview of Azure and the services it offers, in the next chapter, we will dive into some of the core technologies and solutions available to organizations.

Helpful links

- If you'd like to learn more about Azure architecture, you can review the documentation on resources and architectures here: https://docs.microsoft.com/azure/architecture/framework/

- To try Azure using 12 months of free services, go here: https://azure.microsoft.com/free/services/virtual-machines/

Automation and governance in Azure

In the last chapter, we saw how infrastructure in the cloud works from not only a native but also a hybrid perspective. We will now take a look at how we can build resources in Azure. You can now go to the Azure portal and create any Azure resource; however, this can be very cumbersome without automation. Automation in Azure is accomplished through Azure DevOps and **Azure Resource Manager (ARM)** templates. We are strictly sticking to out-of-the-box Microsoft solutions, but there are quite a few other deployment and development tools available that can help you accomplish automation tasks. Once you've deployed your resources, you need to ensure they're secure.

In this chapter, we are going to cover the following:

- Azure DevOps and why it is important
- ARM templates and the different ways they can be used
- Fundamentals and best practices of deploying Azure **Infrastructure as Code (IaC)**
- Benefits and best practices for identity and access control in Azure
- Azure governance

Before we dive into how you can accomplish automation and what it means to have IaC, let's get an overview of Azure DevOps and ARM templates to build a foundation for this automation approach.

Azure DevOps

While this chapter isn't about Azure DevOps, it's a good idea to begin with a fundamental understanding of what it brings to the table. Azure DevOps is both a developer tool and a business tool, as it can be the source of truth for your code base and a backlog of items that code needs to accomplish. Let's look at some of the options that it brings to the table, from which you can pick and choose:

- Azure Repos allows you to either create a Git repository or Team Foundation Version Control to store your development source control.
- Azure Pipelines, one of the critical processes we will use in this chapter for the artifacts we create, provides build and release services for **continuous integration and delivery (CI/CD)** of your apps.
- Azure Boards helps deliver a product backlog to plan and track work, code defects, and other issues that may arise during your software development.
- Azure Test Plans allows you to test the code within your repository and enables you to perform manual and exploratory testing, along with continuous testing.
- Azure Artifacts provides the elements needed for your code to be packaged and deployed, such as NuGet resources usually shared with your CI/CD pipelines.

As you can see, Azure DevOps is Microsoft's tool for deploying and managing applications within Azure as part of the release management process. To learn more about Azure DevOps, you can head over to the documentation at https://docs. microsoft.com/azure/devops/user-guide/what-is-azure-devops?view=azure-devops.

> **Note**
>
> Azure DevOps is available for free with a five-user license, so feel free to have a look and explore how deployment in Azure works. Head to https://azure.microsoft. com/services/devops/ to create your free account.

Now that we've discussed some of the toolings at a very high level, let's take a look at ARM templates.

ARM templates

ARM templates are how your infrastructure is represented as code. ARM templates help teams take a more agile approach to deploying infrastructure in the cloud; it is no longer necessary to click deploy within the Azure portal to create your infrastructure. An ARM template is a mixture of a JSON file representing the configuration of your infrastructure and a PowerShell script to execute that template and create the infrastructure.

The real benefit of using the ARM template system is that it allows you to have declarative syntax. That means you can deploy a virtual machine and create the networking infrastructure that goes around it. Templates end up providing a process that can be run repeatedly in a very consistent manner. They manage the desired state of the infrastructure, meaning a template becomes the source of truth for those infrastructure resources. If you make changes to your infrastructure, you should do that through the templates.

The template deployment process can't be accomplished without orchestrating how the template process needs to run and what order it needs to run in. It is also useful to break these files into smaller chunks and allow them to be linked together or reused in different fashions with other templates. This can help with understanding and controlling your infrastructure while making it repeatable and stable. ARM templates are used in CI/CD pipelines and code deployment to build a suite of applications within the organization.

The following JSON file shows you how ARM templates are structured:

```
{
    "$schema": "https://schema.management.azure.com/schemas/2015-01-01/
deploymentTemplate.json#",

    "contentVersion": "1.0.0.0",

    "parameters": {},

    "variables": {},

    "resources": [],

    "outputs": {}
}
```

As you can see, there are several important parts: parameters, variables, resources, and outputs. Let's discuss each of them briefly:

- ARM template files should be parameterized, and there is a separate file for the parameters that maps to the **parameters** list in the JSON template file.

- The **variables** portion is for variables used within this file. Variables are generally used for creating naming functions to help generate a naming convention that is already structured, and for building that structure to use input parameters to make the name.

- The **resources** section is where all of the resources that you're trying to deploy with this ARM template are represented; these can range from virtual machines to websites.

- Finally, the **outputs** section is anything you want to pass out of your ARM template to be used elsewhere, such as an SQL Server name, before running your SQL scripts.

Three files are created when you create an ARM template file within Visual Studio:

- The first is the JSON file, which is the template that is represented in the preceding code.

- The second is the JSON parameter input file, a file that can be changed with every deployment to match the environment you want to deploy.

- The third is the PowerShell script used to execute the template. The PowerShell script accepts the resource group's inputs, the ARM template file, and the parameters file.

That was a quick overview of Azure DevOps and the files that are created when you create an ARM template. Let's see how Azure resources are deployed using these ARM templates.

Deploying Azure IaC

As we saw in the ARM *templates* section, we want to use ARM templates to deploy our infrastructure, as it fits in nicely with our CI/CD process. There are several ways to approach building out these templates.

One way is to create a monolithic template that contains all of the resources that you want to deploy. To make things a little more modular, you could use a nested template structure. Alternatively, you may want to take a more decoupled approach and create smaller templates that you can link together, making a highly usable and repeatable structure.

Let's take a look at each of these methods, starting with the monolithic view:

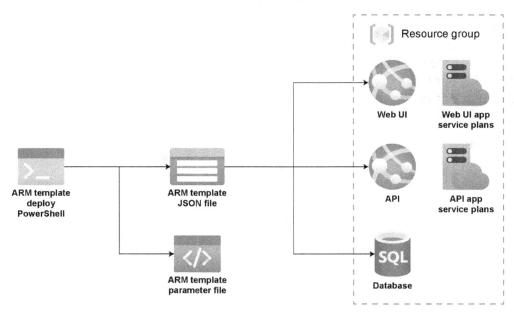

Figure 2.1: Monolithic ARM template

As you can see in *Figure 2.1*, a monolithic ARM template deploys a UI front end with an API middle tier connected to the SQL database. In this process, we need to build out all of the dependencies within the JSON template. The SQL database is deployed before the API middle tier to use the connection string in the API application configuration. You would then deploy the UI layer with the API URL being used in the UI application configuration. The chaining of the deployment can work not only for deploying code but also helping with the configuration.

Alternatively, you could implement a nested template arrangement:

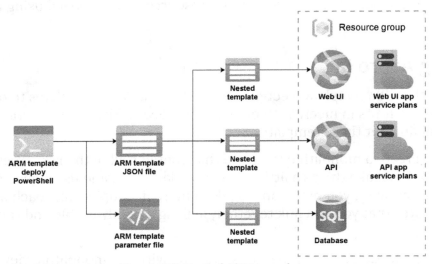

Figure 2.2: Nested ARM templates

As you can see, this is similar to the structure in *Figure 2.1*. However, the templates within this structure are nested in separate file sections. This means that each template owns the resource within it that it's trying to deploy. This structure is similar to breaking out your C# code into manageable methods and actions. This follows the same deployment process as discussed in the monolithic scenario, but the files are nested.

The final structure is linked ARM templates:

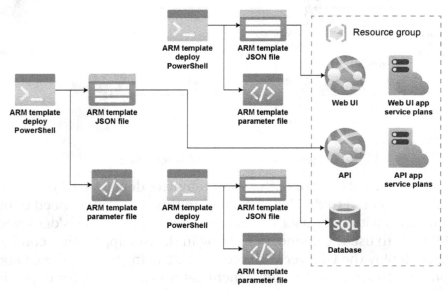

Figure 2.3: Linked ARM templates

As you can see, the templates are initially separate and decoupled from each other, and then we link them together in our release pipeline. Linked templates are similar to nested templates, except the files are external to the template and system, whereas the nested templates are included in the same scope as the parent template. This helps with reusability down the line, because the templates are separate files that can be linked to other deployment files.

We should note that with linked or nested templates, the deployment mode can only be set to **Incremental**. However, the main template can be deployed in **Complete** mode, so if the linked or nested templates target the same resource group, that combined deployment will be evaluated for a complete deployment; otherwise, it will deploy incrementally. To learn more about ARM deployment modes, visit https://docs.microsoft.com/azure/azure-resource-manager/templates/deployment-modes.

We've seen different ways of using these ARM templates to automate the deployment of infrastructure; now we turn to the benefits of doing so.

Benefits of Azure IaC

The main benefit of using IaC is automating the creation, updating, and configuration of different resources and environments. Automation takes the human element out of the picture while adding certain key benefits:

- The ability to schedule automated deployments, helping your operations staff work fewer hours.
- The ability to smoke test your automated deployments.
- The ability to create a repeatable process.
- Pure self-healing applications can be achieved.
- The ability to roll back changes.
- Resource Manager helps in tagging resources.
- ARM takes care of the dependencies of resources in the resource group.

Automation at any point is the real key to using a platform like Azure, whether that is automating your infrastructure deployments or testing to ensure the stability of your production deployments.

In order to reap these benefits, we need to learn how to use ARM templates effectively.

Best practices

We want to take a quick look at some best practices to optimize ARM templates. But first, let's start by understanding what some of our template limits are.

Overall, the template can only be a maximum of 4 MB, and each parameter file is limited to 64 KB. You can only have 256 parameters, with 256 variables, containing 800 resources, 64 output values, and 24,576 characters in a template expression. As we've discussed, you can exceed some of these limits by using nested templates if your template gets too big, but Microsoft recommends that you use linked templates to help avoid these limits. In the following sections, we discuss some best practices for each component within an ARM template.

Parameters

The ARM template system within Azure DevOps resolves parameter values before deployment operations and allows you to reuse the template for different environments. It is essential to point out that each parameter must have a set data type value. You can find a list of these data types at https://docs.microsoft.com/azure/azure-resource-manager/templates/template-syntax#data-types.

Best practices

Microsoft recommends the following best practices for parameters:

- It's best to minimize the use of parameters. As we pointed out at the beginning of the chapter, you should use variables for properties and only use parameters for the things you need to input.

- It is recommended that you use camel casing for parameter names.

- It is also recommended that you describe each parameter, so when other developers use the template they know what the parameters are.

- Ensure that you use parameters for those settings that may change when the environment changes, such as capacity or app service names.

- Ensure you name your parameters to make them easily identifiable.

- Provide default values for parameters; this involves providing the smallest virtual machine skew size so non-production environments use smaller resources and other developers that use the template have a basic starting point.

- If you need to specify optional parameters, avoid using empty strings as the default value and instead use a literal value. This helps to provide a naming structure for users of the template.

- Try to use allowed values as little as possible, as these may change over time and can become difficult to update in your scripts.

- Always use parameters for usernames and passwords or secrets to be set for each environment and not hardcoded in the template. You should also use a secure string for all passwords and secrets.

- When you need to set a location for the resource you're deploying, set the default value to `resourcegroup().location` so the location value is set correctly within the resource group.

As you can see, parameters are very useful in the ARM template process because they allow us to be flexible with the environments we're trying to deploy. Remember to keep these templates as simple as possible with the applications or microservices you're trying to deploy.

Variables

Variables are also resolved before starting the deployment, and the resource manager replaces the variable with its determined value. Variables are useful in deriving complex naming within your template and allow you to only pass in the required parameters.

An example of this is an organization that uses a customer ID and depends on this for its naming convention to keep all deployed resources in Azure unique to that customer ID. In this case, you would create the customer ID as a parameter and then develop variables to generate names using your naming standard. You can find a list of acceptable data types for variables at https://docs.microsoft.com/azure/azure-resource-manager/templates/template-syntax#data-types.

Best practices

Microsoft recommends the following best practices for variables:

- Remember to remove unused variables and files as they can be confusing.

- Use camel casing for your variable names.

- Use variables for values that you need more than once within your template.

- Variable names must be unique.

- For repeatable patterns of JSON objects, use the copy loop in variables.

Resources

The resources section of the ARM templates is reserved for resources that will be deployed or updated. ARM templates generally help derive the desired state of the resources within Azure. When changing Azure infrastructure, it is always a good practice to change your template first and then re-run it to change your Azure resources. All too often, organizations make changes on the portal but forget to change their ARM template, and then the next time they deploy these resources, they are deployed into the wrong state.

Best practices

Microsoft recommends the following best practices for resources:

- Add comments to your resources so that others know their purpose.
- Remember that there are quite a few resources that require unique names, so never hardcode your resource names.
- When you add a password to a custom script extension, use the CommandToExecute property in the protected settings of Azure Resource Manager.

We now have a fundamental understanding of the elements within an ARM template. Our next focus will be identity and access control once your resources have been deployed.

Identity and access control

Before we dive in, it is good to realize that there is a shared responsibility between you and the cloud provider when it comes to security and securing your resources. It is essential to understand where your responsibility stops and the cloud provider steps in. Let's take a quick look at shared responsibility in Azure, as you can see in *Figure 2.4*:

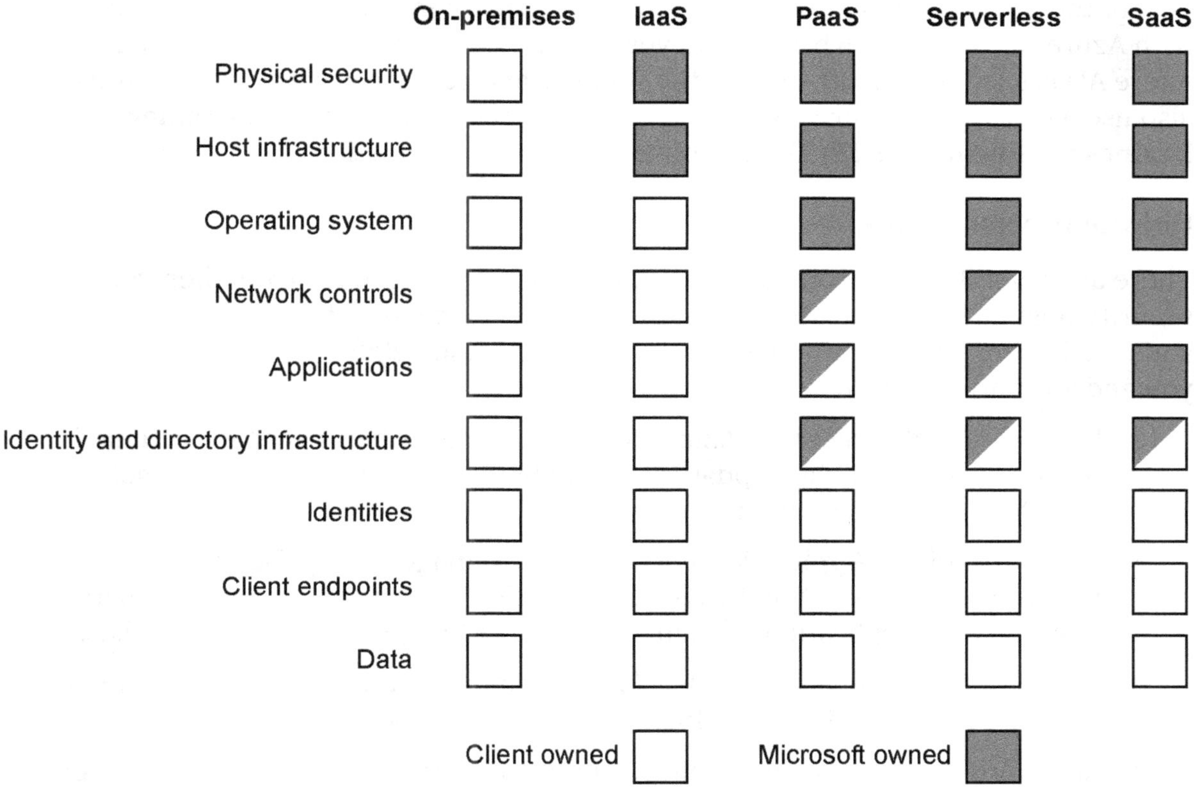

	On-premises	IaaS	PaaS	Serverless	SaaS
Physical security	Client	Microsoft	Microsoft	Microsoft	Microsoft
Host infrastructure	Client	Microsoft	Microsoft	Microsoft	Microsoft
Operating system	Client	Client	Microsoft	Microsoft	Microsoft
Network controls	Client	Client	Shared	Shared	Microsoft
Applications	Client	Client	Shared	Shared	Microsoft
Identity and directory infrastructure	Client	Client	Shared	Shared	Shared
Identities	Client	Client	Client	Client	Client
Client endpoints	Client	Client	Client	Client	Client
Data	Client	Client	Client	Client	Client

Client owned ☐   Microsoft owned ■

Figure 2.4: Azure shared responsibility

Depending on the type of service you choose in Azure, your responsibilities will vary, as will those of the cloud provider. Note the differences between on-premises resources and the various Azure options. You can see that, regardless of any new responsibilities, you will always retain the responsibility for endpoints, account management, accounts, and data repositories you create in the cloud.

With Azure being a public-facing resource, security is at the forefront of its development. There is a wide range of tools and advisors within Azure that help you take advantage of Azure's different security tools and capabilities.

What are the security benefits of Azure?

Organizations' on-premises security groups only have limited resources (team members and tools) to view exploits and attackers. One of the benefits of using a platform like Azure is that you can offload those responsibilities to the provider in the cloud and gain a more efficient and intelligent approach to your organization's threat plane without the need to bring in physical resources.

One of the most significant benefits of Azure is that when you create a tenant, it comes with Azure AD behind it, which allows you to start from a security perspective in Azure. Azure AD is used to lock down all of the services and resources within Azure. You can also use Azure AD to secure your applications or create **Business to Customer** (**B2C**) or **Business to Business** (**B2B**) ADs to house your client information.

Helpful tools in Azure

There are several security and audit solutions built into Azure to strengthen your security posture, which can be viewed in the Azure portal through the security and audit dashboard from your home screen. Here are some helpful tools in Azure to assist you and your organization:

- We mentioned earlier that Azure Resource Manager helps keep everything in one place for deploying, updating, and deleting resources within your solution to support coordinated operations.

- Azure also offers **Application Performance Management** (**APM**), which is referred to as Application Insights. Application Insights gives you the ability to monitor your applications within Azure and detect performance anomalies.

- Azure Monitor allows you to visualize your infrastructure for your activity log and the individual diagnostic logs for your Azure resources.

- Azure Advisor is like a personalized cloud concierge to help you optimize your cloud resources. This service can help detect security and performance issues within your applications.

- Azure Security Center helps prevent, detect, and respond to different threat planes for your applications within Azure. It helps provide security monitoring and policy management across all of your other subscriptions.

Best practices

Here are some Azure security best practices recommended by Microsoft:

- Use Azure AD for central security control and identity management in Azure. This will make management and integration more streamlined.

- Try to keep your Azure AD instances to a single source of truth.

- If you have an on-premises AD, it is recommended that you integrate it with Azure AD, using Azure AD Connect for a single sign-on experience.

- If you use Azure AD Connect to sync your on-premises AD with Azure, turn on password hash synchronization in case the main resource goes offline or is deprecated.

- Remember that you can use Azure AD for authentication in your new applications, and this can be accomplished through Azure AD directly, B2B, or B2C.

- Use management groups to control your access to subscriptions. This helps with centralized management, over needing to worry about **Azure Identity and Access Management (IAM)** in each subscription.

- Use Conditional Access for your support personnel so that they can elevate their permissions when needed in Azure, rather than having access all the time.

- Block legacy protocols that aren't used to stop attack planes.

- It is recommended you use self-service password resets for your users if you're using Azure AD for your applications and you want to ensure you monitor this process.

- If you are using Azure AD Connect, ensure that your cloud policies match your on-premises policies.

- Enable multi-factor authentication for your organization if possible.

- If you wish to provide built-in roles in Azure, ensure that you maintain role-based access over rule-based access, as rule-based access can be very cumbersome to manage in the long run.

- Ensure that you give the least privileged access to those logging in to Azure, so that when an account is compromised, its access is limited.

- Define at least two emergency access accounts, just in case the organization admin operation team members' accounts get compromised.

- Ensure you control locations where resources are created for your organization if you also wish to monitor suspicious activities within your Azure AD tenant actively, as there may be data restrictions in different regions.

- Ensure you use authentication and authorization for your storage accounts.

- Review and apply improvements to your best practices over time.

Now that we have an understanding of security in Azure, we can review how to use Azure governance.

Azure governance

Azure governance is defined as the processes and mechanisms that are used to maintain control of your applications and resources in Azure. They are the strategic priorities involved in planning your initiatives within your organization. Azure governance is provided in two services: Azure Policy and, as we will discuss in *Chapter 7, Offers, support, resources, and tips to optimize cost in Azure*, Azure Cost Management.

The purpose of Azure Policy is to organize your operational standards and to assess your compliance. Azure Policy regulates compliance, security, costs, and management while implementing governance for consistency of your resources. Everything that we see in Azure is governed by these policies, which contain business rules in JSON format and policy definitions. The schema for these policy definitions can be found at https://schema.management.azure.com/schemas/2019-09-01/policyDefinition.json.

What are the benefits?

Azure governance and Azure Policy help with building and scaling your applications while maintaining a level of control. This helps create guardrails and assists with deploying entirely governed environments throughout your organization's subscription using Azure Blueprints. As we will discuss in *Chapter 7, Offers, support, resources, and tips to optimize cost in Azure*, it also assists in managing costs by providing insights into your spending to maximize your cloud investment. In addition, Azure governance offers the following benefits:

- Helps with audit and enforcement of your policies for any Azure service
- Helps encourage accountability throughout the organization while monitoring spending
- Creates compliant environments, including resources, policies, and access control
- Helps ensure compliance with external regulations via built-in compliance controls

In the following sections, we'll look in detail at some of the features and services available via Azure governance, namely Azure management groups, Azure Policy, Azure Blueprints, Azure Graph, and Azure Cost Management and Billing.

Azure management groups

Azure management groups help manage your Azure subscriptions by grouping them and taking actions against those groups. They allow you to define security, policies, and typical deployments via blueprints. They help create a hierarchical view of your organization so that you can efficiently manage your subscriptions and resources:

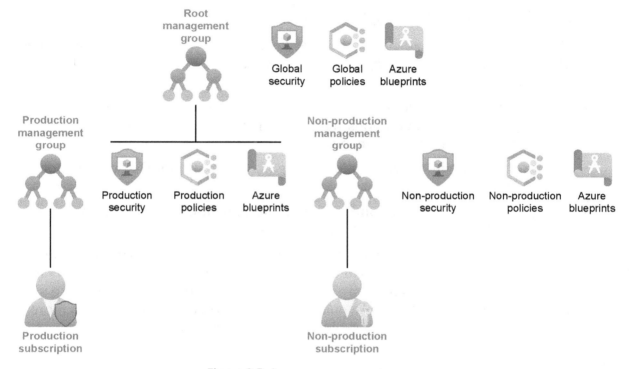

Figure 2.5: Azure management groups

As you can see in *Figure* 2.5, there is a simple separation between production and non-production. We should understand from this illustration that each management group has a root or parent that every hierarchical structure inherits from. You can create a global structure at the root, or you can create a production or non-production policy in either one of the branches.

Azure Policy

Azure Policy was put in place to help enforce asset compliance in organizational standards within Azure. Common uses of Azure Policy are implementing governance for resources consistency, security, costs, and management. Like everything in Azure, Azure policies are in JSON format, and you add business rules for policy definitions to help simplify the management of these rules.

Azure policies can be applied to Azure resources in different life cycles or during an ongoing compliance evaluation. These can be used as a control mechanism to deny changes or to log them. The difference between Azure Policy and Azure **rule-based access control (RBAC)** is that your policy does not restrict Azure actions. This means a combination of Azure RBAC and Azure policy provides the full scope of security in Azure.

The following best practices are recommended by Microsoft:

- When you use Azure policies, it's always good to start with an audit policy rather than a denial policy, as setting a denial policy may hinder automation tasks when creating the resources.

- When creating your definitions, consider your organizational hierarchies. Creating higher-level definitions such as at the management group or subscription level is recommended.

- Create and assign initiative definitions or policy sets even for the smallest policy definitions.

- It is good to remember that once an initial assignment is evaluated, all policies within that initiative are evaluated as well.

- You should think about using policies to help control your infrastructure, like requiring antivirus to be installed on all virtual machines or not allowing specific sizes of virtual machines to be created in a non-production environment. To gain a better understanding of Azure policy definition structure, you can learn more at https://docs.microsoft.com/azure/governance/policy/concepts/definition-structure.

Azure Blueprints

Azure Blueprints enables technology groups to develop a repeatable set of Azure resources that support an organization's patterns, requirements, and standards. Blueprints are a great way to orchestrate the deployment of various resources, such as role assignments, policies, ARM templates, and resource groups. Azure Blueprints is an extension of ARM templates, which are designed to help with environment setup, and Azure Blueprints uses templates to accomplish this goal.

Azure Resource Graph

Azure Resource Graph was created to extend ARM templates' capabilities to help explore resources even across subscriptions. Azure Resource Graph queries allow you to search for complex results from resources that have been deployed in Azure. Azure Resource Graph is the query system that supports the search in Azure. The query language is based on the Kusto Query Language, which is also used by Azure Data Explorer, so it may be new to you and take a little bit of getting used to.

You need the appropriate rights in Azure RBAC to see the resources—this is the read permission. If you don't get any results returned in Azure when you use Azure Resource Graph, check your read permission first.

Azure Resource Graph is free to use, but it's throttled to ensure the best experience for everyone.

Azure Cost Management and Billing

Azure Cost Management and Billing was created to help analyze, manage, and optimize the costs of your workloads in Azure. It was introduced to help businesses reduce their risk of potential waste and inefficiencies as they migrate to the cloud. Azure Cost Management and Billing does the following:

- Assists in paying your bills
- Generates monthly invoices containing cost and usage data that can be downloaded
- Sets spending thresholds
- Analyzes your costs proactively
- Identifies opportunities to optimize spending for your workloads in Azure

We will cover this in more depth in *Chapter 7, Offers, support, resources, and tips to optimize cost in Azure*, since individuals and organizations manage cost streams differently in Azure.

Summary

As you can see from this chapter, Azure and Azure DevOps have significant synergies. It is essential to create a repeatable, stable way to deploy your code and infrastructure to the cloud. While learning how to deploy this infrastructure to the cloud, we needed to understand why ARM templates are used. This led us to discuss some of the fundamentals and best practices around deploying this infrastructure to leverage our code or applications. We looked at exactly how we need to secure resources and our applications through identity and access control. This brought us to understanding how to create governance in Azure to ensure consistency and compliance.

Azure creates a lot of frameworks that allow you to digest the things that you need to leverage to ensure your applications are secure and complete. It is also good to remember that as you approach Azure, you should approach it from an automation perspective. An organization should develop compliance, Azure governance, and best practices that work across the organization while minimizing its business risk.

Now that we have built a foundation, in the next chapter, we will move on to how we can modernize applications.

Important links

- *Azure Cost Management and Billing*: https://docs.microsoft.com/azure/cost-management-billing/cost-management-billing-overview
- *Azure Resource Graph*: https://docs.microsoft.com/azure/governance/resource-graph/
- *Azure Blueprints*: https://docs.microsoft.com/azure/governance/blueprints/overview
- *Azure Policy*: https://docs.microsoft.com/azure/governance/policy/overview#policy-definition
- *Azure Policy*: https://docs.microsoft.com/azure/governance/policy/concepts/definition-structure
- *Azure management groups*: https://docs.microsoft.com/azure/governance/management-groups/
- *ARM Template Toolkit*: https://docs.microsoft.com/azure/azure-resource-manager/templates/test-toolkit

- ARM *Template Structure*: https://docs.microsoft.com/azure/azure-resource-manager/templates/template-syntax
- ARM *Template Recommendations*: https://docs.microsoft.com/azure/azure-resource-manager/templates/templates-cloud-consistency
- *Azure Security Center*: https://azure.microsoft.com/services/security-center/
- *Azure Security Best Practices*: https://azure.microsoft.com/blog/azure-storage-support-for-azure-ad-based-access-control-now-generally-available/
- *Resources Link*: https://docs.microsoft.com/azure/azure-resource-manager/templates/template-syntax#resources
- ARM *Template Best Practices*: https://docs.microsoft.com/azure/azure-resource-manager/templates/template-best-practices?WT.mc_id=azuredevops-azuredevops-jagord
- *Variables*: https://docs.microsoft.com/azure/azure-resource-manager/templates/template-variables
- *Parameters*: https://docs.microsoft.com/azure/azure-resource-manager/templates/template-parameters

3

Modernizing with hybrid cloud and multicloud

Business environments are becoming increasingly complex. Many software applications now run on different systems located on-premises, off-premises, in multiple clouds, and at the edge. Proper planning, implementation, and management of these diverse environments are critical factors in helping your users and your organization to make the most of them. This chapter focuses on the role of Azure hybrid and multicloud solutions—in particular, Azure Arc and the Azure Stack family of solutions.

By the end of this chapter, you will be able to do the following:

- Understand what hybrid cloud, multicloud, and edge computing are
- Discuss what makes a hybrid and multicloud strategy successful
- Explain Azure Arc enabled servers, data services, and Kubernetes app management
- Present the Azure Stack portfolio and explain how it can modernize your datacenter

This chapter will teach you about Azure Arc connectivity requirements and approaches for creating Azure Arc enabled data services on Kubernetes clusters and managed Kubernetes services. You will also learn how to include the Azure Stack family of solutions in your architecture and manage Azure Stack Hub. In the first section, we'll kick things off with some terminology.

What is hybrid cloud, multicloud, and edge computing?

We have already introduced hybrid cloud, multicloud, and edge computing in *Chapter 1, Introduction*, but here's a quick recap:

- A **hybrid cloud** approach involves the combination of cloud and non-cloud resources (such as an on-premises datacenter). This can offer greater flexibility, more deployment options, scalability, and operational consistency.
- A **multicloud** approach involves leveraging cloud computing services from multiple service providers. This can provide increased flexibility and better risk mitigation, as you can choose a combination of services and regional providers that best fits the needs of your organization.
- An **edge computing** approach harnesses the computing power of the cloud in devices close to the endpoints where data is created and consumed. These endpoints could include process control systems on production lines, building surveillance systems, and sensors and actuators at remote sites.

Which of these computing environments you choose to use depends on a variety of different factors. For instance, a hybrid cloud approach that includes on-premises systems might be the best choice for your organization to meet regulatory and data sovereignty requirements while enhancing resilience and business continuity.

By contrast, if your organization wants to reduce response latency and ensure offline availability in the field, edge computing, with its **virtual machines** (**VMs**), containers, and data services per device, could be essential.

It is important to keep in mind that hybrid cloud, multicloud, and edge computing may also add operational complexity. For example, multicloud necessitates dealing with different cloud environments across different service providers, which poses an additional challenge for any organization already facing a shortage of cloud talent. In this case, a solution like Azure can help because it is designed to facilitate the management of multicloud as well as hybrid cloud and edge computing.

Now that we have discussed terminology, let's talk about strategy.

What makes a hybrid and multicloud strategy successful?

Enabling your teams to use the technology that fits their needs while providing security and governance across your organization's locations is a crucial goal if you want to use hybrid and multicloud solutions successfully. Besides improving your end users' experience, such an environment also enables unified operations for application development and IT management. Your developers can build apps with the same tools, APIs, and deployment across your organization. Your management, security, and governance can be made homogeneous for all your business locations. Consistent choices of databases and computing stacks across locations let you quickly move data and workloads as you need.

Overall, Azure hybrid and multicloud solutions and products let you optimize:

- **User app experience**: Create a consistent app experience across your entire cloud, on-premises, and off-premises estate.
- **Data services**: Data migration, management, and analysis can be made seamless by running Azure data services wherever you need them.
- **IT management**: Unify the management, governance, and security of your IT resources with centralized control across all your organizational domains.
- **Security and threat protection**: Use one control plane across your digital estate for security and advanced threat protection for all your workloads.
- **Identity and user access**: Use a unified platform for identity and access management for a seamless, single sign-on user experience globally.
- **Networking**: Connect distributed workloads and locations securely across your organization by making Azure an extension of your current network.

Azure Arc and Azure Stack are two components of Azure hybrid and multicloud solutions and products. Azure Arc provides simplified management of your deployments and extends Azure application services and data across multicloud, datacenters, and the edge. Azure Stack, through its three offerings and ability to optionally utilize Azure Arc, provides:

- A cloud-native integrated system that brings Azure cloud services on-premises, targeting disconnected scenarios.
- A **hyperconverged infrastructure (HCI)** that modernizes datacenters by refreshing virtualization hosts using modern software-defined storage and networking combined with Hyper-V for compute.
- Cloud-managed appliances/edge devices for running edge-computing workloads.

Together, they enable you to:

- Use a single control plane to manage, govern, and secure servers, Kubernetes clusters, and apps seamlessly across multicloud, on-premises, and the edge.
- Keep your Azure services up to date on any infrastructure, along with automation, unified management, and security.
- Modernize applications on-premises or at the edge with cloud-native technologies.
- Combine virtualized apps with a local HCI that can easily add Azure services to the offering for optimal price performance.
- Extend Azure compute, storage, and AI to the IoT and other edge devices, and run machine learning and advanced analytics at the edge for real-time insights.

> **Note**
>
> This chapter focuses on Azure Arc and Azure Stack planning, implementation, and best practices. Further information on Azure security and related topics is available at https://azure.microsoft.com/services/security-center/ and https://docs.microsoft.com/azure/?product=all.
>
> Microsoft also offers examples of solution architecture for building hybrid and multicloud solutions and a flexible migration path. Here is one for managing configurations for Azure Arc enabled servers: https://docs.microsoft.com/azure/architecture/hybrid/azure-arc-hybrid-config.

Hopefully, these first sections have given you a good introductory foundation. In the next section, we'll take a deeper dive into Azure Arc.

Azure Arc overview

As cloud, non-cloud, on-premises, and off-premises solutions continue to expand at ever-increasing rates, organizations must manage increasingly complex IT environments. Multiple clouds and environments means multiple disparate management tools and learning curves. As an additional challenge, existing tools may not adequately support such new cloud-native versions of operational models like DevOps and ITOps.

Azure Arc offers a solution to simplify the management and governance of these environments. With Azure Arc, you can deploy Azure services and extend Azure management anywhere. It is a single platform (single pane of glass) for the consistent management of multicloud, on-premises, and edge resources, each projected via Azure Arc into **Azure Resource Manager** (**ARM**). This lets you manage VMs, Kubernetes clusters, and databases as if they were running in Azure, using Azure services and management tools.

The key benefits of Azure Arc are:

- Single-pane-of-glass visibility of your operations and compliance. Use the Azure portal to manage, govern, and secure a wide range of resources across Windows, Linux, SQL Server, and Kubernetes spanning datacenters, the edge, and multicloud.
- The ability to architect and design hybrid applications at scale where components are distributed across public cloud services, private clouds, datacenters, and edge locations without sacrificing central visibility and control. Centrally code and deploy applications confidently to any Kubernetes distribution in any location. Accelerate development through standardized deployment, configuration, security, and observability.
- The extension of Azure application services and data to all your locations. Utilize Azure data and application services to implement cloud practices and automation to consistently deploy scalable, up-to-date Azure Arc enabled services.

With Azure Arc, you can:

- Continue with and extend the application of the processes and services administered by your IT department for your business (ITOps) across all these environments.
- Deploy DevOps to support new cloud-native patterns across all environments, including cloud and non-cloud.
- Manage and govern Windows and Linux servers, including physical and virtual machines, inside and outside Azure.
- Manage and govern Kubernetes clusters at scale and with source control for the consistent deployment and configuration of applications.
- Manage Azure SQL Database and PostgreSQL Hyperscale services (Azure Data Services) with on-demand scaling backed by automated patching, upgrades, and security across your organizational domains inside and outside the cloud.

Specifically, Azure Arc key features enable:

- Consistent server inventory, management, governance, and security across your environments.
- Monitoring, security, and updates of your servers by configuring Azure VM extensions for Azure management services.
- Management and scaling of Kubernetes clusters using any CNFC-conformant Kubernetes distribution using DevOps techniques.
- GitOps-based management with **Configuration as Code** (**CaC**) for the deployment of applications and configuration across multiple clusters, directly from source control.
- The use of Azure Policy for automated (zero-touch) compliance and configuration for your Kubernetes clusters.
- Deployment of Azure SQL Managed Instance and Azure Database for PostgreSQL Hyperscale (Azure data services) in any Kubernetes environment, with Azure-level upgrade/update, security, monitoring, and leveraging Azure SQL high availability features and PostgreSQL Hyperscale disconnected scenario support.
- A unified view of your Azure Arc enabled assets, wherever they are, whether via the Azure portal, CLI, PowerShell, or REST API.

> **Note**
>
> To run Kubernetes workloads on Azure Stack HCI, customers can deploy Azure Kubernetes Service on Azure Stack HCI, which is a service specially architected for Azure Stack HCI.

Now we have an overview of what Azure Arc can help you accomplish, let's go into some detail. We will first go over Azure Arc enabled infrastructure, starting with how it facilitates the management of Kubernetes at scale. Later in this section, we will cover Azure Arc enabled data services that make it possible to run Azure data services on-premises, in other public clouds, and at the edge.

Kubernetes app management at scale

The advantages of containers, including their portability, efficiency, and scalability, have made them popular for deploying business applications. New applications can be created as microservices in Kubernetes clusters. Legacy applications can be modernized by rewriting them as containers. However, with these possibilities comes the need for management of Kubernetes at scale.

Azure Arc addresses this need, letting organizations deploy applications quickly to any Kubernetes cluster across multiple locations under robust management policies. This capability is enabled through **Azure Kubernetes Configuration Management (AKCM)**, an Azure service that delivers configuration management and application deployment from Azure, using GitOps. Continuous, compliant deployment can be assured by linking application policies to specific GitHub repositories. With this capability, cluster admins can declare their cluster configuration and applications in Git. The development teams can then use pull requests and the tools they are familiar with (existing DevOps pipelines, Git, Kubernetes manifests, and Helm charts, for example) to easily deploy applications onto Azure Arc enabled Kubernetes clusters and make updates in production. The GitOps agents listen to changes and facilitate automated rollbacks if these changes result in system divergence from the source of truth.

For example, a retailing enterprise with many outlets can migrate its in-store applications to Kubernetes clusters of containers. Azure Arc enables the uniform deployment, configuration, and management of these containerized applications to multiple locations. New outlets can receive specific sets of applications with centralized control of compliance and configuration. The IT team can also monitor, secure, and change configurations and applications in all outlets while leveraging policies to secure network connections and prevent misconfigurations.

Azure Kubernetes Service (AKS) can be used to monitor Kubernetes cluster health, carrying out maintenance, mounting storage volumes, and activating nodes for specific tasks (for example, using GPU-enabled nodes for parallel processing). Azure Arc and Azure Policy then give the retailer's IT team a unified view in the Azure portal of all clusters in all outlets. You can also run Azure Arc enabled data services on Azure Stack HCI. This is discussed later in the chapter.

Azure Arc enabled servers

With Azure Arc enabled servers, you can apply native Azure VM levels of management to Windows and Linux machines located outside Azure, for example, on an enterprise network or as part of services from other cloud providers. These non-Azure resources can be connected to Azure and managed as Azure resources. Each connected hybrid machine is then included in a resource group with a machine-level resource ID. It can also be managed using standard Azure functionality such as Azure Policy, RBAC, and tags.

For example, service providers managing multiple customer environments can extend native Azure management functionality via Azure Arc, such as Azure Lighthouse, which lets service providers sign in to their tenant so they can manage subscriptions and resource groups as delegated by customers.

Azure Arc's control plane functionality provides:

- Management groups and tags for organizing your resources.
- The use of Azure Resource Graph for indexing and searching.
- The use of Azure **Role-Based Access and Control** (**RBAC**) for security and access control.
- Templates and extensions for environment creation and automation.
- Update management.

Supported scenarios

When you connect your machine to Azure Arc enabled servers, the following benefits are available:

- **Azure Automation Change Tracking and Inventory**: Used for reporting on configuration changes on monitored servers. This can be used for Microsoft services, Windows registry and files, Linux daemons, and your installed software.
- **Azure Automation State Configuration**: Used for simplifying deployment for your non-Azure Windows or Linux machines. You can also use Custom Script Extension for software installation or post-deployment configuration.
- **Update Management**: For managing Windows and Linux server operating system updates. For more information, check out https://docs.microsoft.com/azure/automation/update-management/enable-from-automation-account.
- **Azure Monitor for VMs**: This allows you to monitor guest operating system performance on your connected machine. You can discover application components and monitor their processes and dependencies.
- **Azure Policy guest configurations**: You can assign (and not just audit) Azure Policy guest configurations to your hybrid machine, just like you assign policies to Azure native machines.
- **Azure Defender**: Used for detecting and responding to threats, as well as managing the preventative security and compliance capabilities of your non-Azure servers.

> **Note**
>
> Each hybrid machine to be managed via Azure Arc requires the installation of the Azure Connected Machine agent. For proactive monitoring of the operating system and workloads on the machine, you should also install the Log Analytics agent for Windows and Linux, then manage the device with automation runbooks, Update Management, Azure Security Center, or other suitable services. Log data collected and stored from the hybrid device can then be identified in a Log Analytics workspace via its resource ID or other properties.

Supported regions

Whether you enable Azure Arc enabled servers manually or through running a template script, the natural choice for the location is the Azure region geographically nearest to your device. Your choice of region may also be determined by data residency requirements, given that data is stored in the Azure geography that contains the chosen region.

If there is an outage in the Azure region you select, your connected machine will not be affected. On the other hand, Azure-based management operations may not be able to complete. For a geographically redundant service with multiple devices or locations, connect them to different Azure regions.

Some data is collected and stored from the connected machine. The following is stored in the region where the Azure Arc machine resource was created:

- The computer **fully qualified domain name (FQDN)**
- The name of the computer
- The version of the Connected Machine agent
- The name and version of the operating system

Every 5 minutes, the connected machine sends a heartbeat signal to the service. If these signals cease, the service will change the status of the device in the portal to disconnected within 15 to 30 minutes. When the connected machine agent sends a new heartbeat signal, the status will change back to connected.

For further reading, check out:

- https://docs.microsoft.com/azure/azure-arc/servers/overview
- https://docs.microsoft.com/azure/azure-arc/servers/agent-overview

Azure Arc enabled data services

With Azure Arc and Kubernetes, specified Azure data services are available for you to run on the infrastructure of your choice outside Azure, for example, on-premises, on edge devices, and in other clouds. Azure innovations, scalability, and unified management are then available for data workloads in both connected and disconnected scenarios. The first Azure Arc enabled data services to be made public (first in preview) are SQL Managed Instance and PostgreSQL Hyperscale. It is important to note that you can also run Azure Arc enabled data services on Azure Stack HCI, which is discussed later in this chapter.

Evergreen (always current)

By using Azure Arc enabled data services for on-premises workloads, you can systematically access the latest Azure features and capabilities. You can also configure automatic updates to receive the latest patches and upgrades from Microsoft Container Registry while controlling deployment cadences according to your policy and maximizing system uptime. As a further advantage, you can avoid end-of-support issues for your databases, because Azure Arc enabled data services are a subscription service.

Elasticity

Azure Arc enabled data services offer cloud-like elastic scaling for your on-premises databases, allowing you to meet the requirements of volatile and bursting workloads, as well as real-time data ingestion and queries. No limit is placed on scalability and you can spin up database instances in seconds for sub-second operational response times. Database administrative tasks such as configuring high availability are simplified to the level of a few clicks. Data workloads can dynamically scale according to the capacity needed and without halting applications, with both scale-up and scale-out possibilities for increased read replicas or sharding.

Unified management

You can get a unified view of your data assets deployed with Azure Arc by using tools such as the Azure portal, Azure Data Studio, or Azure Data CLI. You can check on infrastructure capacity and health using logs and telemetry from Kubernetes APIs. Azure Monitor is also available for operational views and insights across all your data estate.

To scale management with resources, Azure Arc automates database management tasks. Out-of-the-box functionality includes monitoring, rapid provisioning, on-demand elastic scaling, patching, high availability setup, and backup and restore. With Azure Arc, databases in your digital domains can also benefit from Azure Backup, Azure Monitor, Azure Policy, Azure RBAC, and Advanced Data Security.

Management in disconnected scenarios

You can access Azure benefits with or without a direct and continuous cloud connection. Many services, including monitoring, automated backup/restore, and self-service provisioning, can run locally whether a direct connection to Azure is available or not. The local Azure Arc data controller offers comprehensive management functionality in your self-hosted environment for provisioning, elastic scale, backup, automated updates, monitoring, and high availability. Connecting directly to Azure adds options for integration with other Azure services such as Azure Monitor, and enables you to use the Azure portal and Azure Resource Manager APIs from anywhere for management of your Azure Arc enabled data services.

Prerequisites for Azure data services

You will need a Kubernetes cluster based on a major Kubernetes distribution for the orchestration of Azure data services on the infrastructure of your choice. You will also need to install the Azure Arc data controller before provisioning Azure data services and using management functionality in your environment.

Should you choose an Arc enabled SQL server or Arc enabled managed instance?

Azure already offers different deployment and management options for hosting SQL Server capability. Support for SQL Server with Azure Arc increases the range of possibilities. Azure Arc possibilities can be compared as follows:

- **Azure Arc enabled SQL Server (currently in preview)**: For SQL servers in your own infrastructure or another public cloud, Azure Arc enabled SQL Server lets you connect these SQL servers to Azure and take advantage of the associated Azure services. There is no impact on the SQL servers from connection and registration with Azure. There is no requirement for data migration or downtime either. With the Azure portal as your central management dashboard, you can manage all your SQL servers. The SQL Server On-Demand Assessment service lets you regularly validate the health of your SQL Server environment, mitigate risks, and enhance performance.
- **Azure Arc enabled SQL Managed Instance (currently in preview)**: Azure Arc enabled SQL Managed Instance is an Azure SQL data service. It can be created anywhere in your infrastructure and has close to 100% compatibility with the most recent SQL Server database engine. With Azure Arc enabled SQL Managed Instance, you can lift and shift your applications to Azure Arc data services while minimizing changes to your applications and databases.

Without leaving your existing infrastructure, you can migrate your existing SQL Server applications to the latest version of the SQL Server engine and get the additional advantage of the PaaS-like integrated management capabilities. These capabilities help you meet compliance criteria such as data sovereignty. To do this, you leverage the Kubernetes platform with Azure data services, for which deployment can be on any infrastructure.

Currently, the following advantages are available:

- Easy creation, removal, and elastic scale-up or down of a managed instance within a minute.
- The platform automatically installs upgrades, updates, and patches to ensure that you run the most recent version of SQL Server.
- Monitoring, high availability, and backup and restore available as integrated management services.

Azure Arc enabled PostgreSQL Hyperscale or Azure Database for PostgreSQL Hyperscale?

These two entities differ from each other in a similar way as Arc enabled SQL Server and Arc enabled SQL Managed Instance do with each other. Azure Database for PostgreSQL Hyperscale is a service in Azure operated by Microsoft and running in Microsoft datacenters. By comparison, Azure Arc enabled PostgreSQL Hyperscale is part of Azure Arc enabled data services and runs on your own infrastructure. However, both entities are based on the hyperscale form factor of the PostgreSQL database powered by the Citus extension.

Connectivity modes

Your Azure Arc enabled data services environment can connect to Azure in different ways, depending on factors like your business policy, government regulations, and the network connections available. With Azure Arc enabled data services, you can choose between the following connectivity modes:

- Directly connected (not supported at the time of writing)
- Indirectly connected
- Never connected (not supported at the time of writing)

The availability of some functionality of Azure Arc enabled data services will depend on the connectivity mode that you choose. The choice of connectivity mode lets you decide how much data is transmitted to Azure and the type of user interaction with the Arc data controller. Let's discuss the differences between direct and indirect connections.

Direct connections

When Azure Arc enabled data services are directly connected to Azure:

- Users can operate the Azure Arc data services via Azure Resource Manager APIs, the Azure CLI, and the Azure portal.
- **Azure Active Directory (Azure AD)** and Azure RBAC are available because of the continuous and direct communication in the directly connected mode.
- Services such as Azure Defender security services, Container Insights, and Azure Backup to blob storage are available in the directly connected mode.

Operations in the directly connected mode are similar to using the Azure portal for services, for example, provisioning/de-provisioning, configuring, and scaling.

Direct connections may be used for:

- Organizations using public clouds such as Azure, AWS, or Google Cloud Platform.
- Edge site locations such as retail stores where internet connectivity is often present and permitted.
- Corporate datacenters, allowing more extensive connectivity to/from the data region of their datacenter and the internet.

Indirect connections

When Azure Arc enabled data services are indirectly connected to Azure:

- Only a read-only view is available in the Azure portal. You can see instances and details of deployments of managed instances and Postgres Hyperscale instances, but you cannot act on them in the Azure portal.
- All actions must be initiated locally using Azure Data Studio, Azure Data CLI (azdata), or Kubernetes-native tools such as kubectl.
- Azure AD and Azure RBAC are not available.
- Services such as Azure Defender security services, Container Insights, and Azure Backup to blob storage are not available.

At the time of writing, only the indirectly connected mode is supported (in preview).

Indirect connections may be used for:

- On-premises datacenters, for example, for finance, healthcare, or government, may disallow connectivity in or out of the data region. This is due to business or regulatory compliance policies, often to avoid the risks of external attacks or data exfiltration.
- Edge site locations, for example, for oil/gas or military field applications, where connectivity to the internet is often unavailable.
- Edge site locations, such as ships, with only intermittent connectivity.

Never connected

In never connected mode, no data can be sent in any way to or from Azure. A use case for this scenario could be a top-secret government facility. This type of air-gapped environment ensures full data isolation.

It is important to note that this mode is not yet supported.

Connectivity requirements

An agent in your environment is always the initiator of communication between your environment and Azure. This is also true for operations initiated by a user in the Azure portal, which become tasks queued up in Azure. An agent in your environment then checks for tasks in the queue by initiating communication with Azure. The agent runs those tasks and reports to Azure on the task status (completion or failure).

Connections available for the indirectly connected mode

There are currently three connections available for the only supported mode in the preview phase, the indirectly connected mode. They are:

- Azure Monitor APIs
- Azure Resource Manager APIs
- **Microsoft Container Registry (MCR)**

All HTTPS connections to Azure and the Microsoft Container Registry are encrypted. They use SSL/TLS and officially signed and verifiable certificates.

Azure Monitor APIs and Azure Resource Manager APIs

Azure Data Studio, Azure Data CLI (azdata), and the Azure CLI connect to the Azure Resource Manager APIs to transmit data to, and receive data from, Azure for certain features.

Currently, all browser HTTPS/443 connections to the Grafana and Kibana dashboards and from Azure Data CLI (azdata) to the data controller API use SSL encryption with self-signed certificates. A feature is planned to make encryption of these SSL connections possible using your own certificates.

Microsoft Container Registry

MCR is the repository for the Azure Arc enabled data services container images. You can pull these images from MCR and push them to a private container registry. You can then configure the data controller deployment process to pull the container images from that private container registry.

Connections to the Kubernetes API server from Azure Data Studio and Azure Data CLI (azdata) use the Kubernetes authentication and encryption that you have put in place. To execute many of the actions concerning Azure Arc enabled data services, users of Azure Data Studio and Azure Data CLI must have an authenticated connection to the Kubernetes API.

Azure Arc data controller creation

Several different approaches are possible for creating Azure Arc enabled data services on Kubernetes clusters and managed Kubernetes services.

You can find a current list of the supported Kubernetes services and distributions here: https://docs.microsoft.com/azure/azure-arc/data/create-data-controller.

This URL will also provide you with requirements for the services and distributions. For example, the minimum supported version of the application, VM size, memory and storage, and connectivity requirements. In addition, you will find the information needed during the controller creation process.

Azure Arc in GitHub

The GitHub repository at https://github.com/microsoft/azure_arc holds different resources to assist you in using Azure Arc enabled servers, Azure Arc enabled SQL Server, Azure Arc enabled Kubernetes, and Azure Arc enabled data services.

We'll take a look at the different guides available that can assist you in Azure Arc adoption.

Azure Arc enabled servers

These deployment scenarios are available to guide you in onboarding different Windows and Linux server deployments to Azure with Azure Arc:

- **General**: Examples that can be used to connect existing Windows or Linux servers to Azure with Azure Arc.
- **Microsoft Azure**: Guides to onboarding an Azure VM as an Azure Arc enabled server.
- Additional guides give information on using Vagrant, **Amazon Web Services (AWS)**, **Google Cloud Platform (GCP)**, and VMware.
- **Azure Arc enabled servers–Day-2 scenarios and use cases**: After server resources have been projected into Azure with Azure Arc to make them Azure Arc enabled servers, these guides show examples of managing these servers as native Azure resources using native Azure management tools such as resource tags, Azure Policy, and Log Analytics.
- **Azure Arc enabled servers–scaled deployment scenarios**: Guides to the scaled onboarding to Azure Arc of VMs in different platforms and existing environments.

Azure Arc enabled SQL Server

These deployment scenarios offer guidance to onboarding deployments of Microsoft SQL Server on different platforms to Azure Arc:

- **Microsoft Azure**: Walkthroughs of onboarding an Azure VM with an SQL Server installation as Azure Arc enabled SQL and as an Azure Arc enabled server. This guide is for demo and testing purposes only (it is not supported).
- **AWS**: Guidance for end-to-end deployment in AWS of Windows Server with SQL Server and onboarding with Terraform to Azure with Azure Arc.
- **GCP**: Guidance for end-to-end deployment in GCP of Windows Server with SQL Server and onboarding with Terraform to Azure with Azure Arc.
- **VMware**: Guidance for end-to-end deployment in VMware vSphere of Windows Server with SQL Server and onboarding with Terraform to Azure with Azure Arc.

Azure Arc enabled Kubernetes

This section has options for Azure Arc enabled Kubernetes, for rapidly spinning up a Kubernetes cluster ready for projection in Azure Arc and management by Azure-native tools:

- **General**: Demonstration by the example of connection to Arc or an existing Kubernetes cluster.
- **AKS**: Example walkthroughs of creating an AKS cluster to simulate a cluster running on-premises. Deployment examples include using Terraform and an ARM template.
- **Amazon Elastic Kubernetes Service (EKS)**: An example using Terraform to deploy an EKS cluster on AWS and connection of this EKS cluster to Azure with Azure Arc.
- **Google Kubernetes Engine (GKE)**: An example using Terraform to deploy a GKE cluster on Google Cloud and the connection of this EKS cluster to Azure with Azure Arc.
- **Rancher k3s**: Deployment examples of Rancher k3s lightweight Kubernetes (for example, for an edge, IoT, or embedded Kubernetes) on an Azure VM or VMware onboarding of the cluster to Azure with Azure Arc.
- **Azure Red Hat OpenShift (ARO) V4**: An example using Terraform to deploy a new ARO cluster with onboarding of the cluster to Azure with Azure Arc.
- **Kubernetes in Docker (kind)**: An example of using kind for the creation of a Kubernetes cluster on your local machine (running as a local Kubernetes cluster using Docker container "nodes") and onboarding of the cluster as an Azure Arc enabled Kubernetes cluster.
- **MicroK8s**: An example walkthrough of using MicroK8s to create a Kubernetes cluster on your local machine and onboarding of the cluster as an Azure Arc enabled Kubernetes cluster.
- **Azure Arc enabled Kubernetes–day-2 scenarios and use cases**: After Kubernetes clusters have been projected into Azure with Azure Arc, these guides show examples of managing these clusters as native Azure resources using native Azure management tools such as Azure Monitor, Azure Policy, and GitOps configurations. For instance, for AKS (works also for GKE), you can deploy GitOps configurations and perform basic and Helm-based GitOps flow on AKS as an Azure Arc connected cluster. You can also apply GitOps configurations on AKS as an Azure Arc connected cluster using Azure Policy for Kubernetes and integrate Azure Monitor for containers with AKS as an Azure Arc connected cluster.

Azure Arc enabled data services (currently in preview)

This section contains deployment options for Azure Arc enabled data services. The goal is the rapid spinning-up of new Kubernetes clusters and deployment of Azure Arc enabled data services that are ready for projection in Azure Arc and management with Azure-native tooling. If you already have a Kubernetes cluster, you can use this information to deploy Azure Arc enabled data services to an existing Kubernetes cluster: https://docs.microsoft.com/azure/azure-arc/data/create-data-controller:

- **Data services on AKS**: Example walkthroughs for creating an AKS cluster with the deployment of Azure Arc data services on the cluster. For example, use of Azure Arc Data Controller Vanilla, Azure SQL Managed Instance, and Azure PostgreSQL Hyperscale Deployment on AKS with an Azure ARM template: https://github.com/microsoft/azure_arc/tree/main/azure_arc_data_jumpstart/aks/arm_template.
- **Data services on AWS Elastic Kubernetes**: Example walkthroughs for the creation of an EKS cluster and the deployment of Azure Arc data services on that cluster.
- **Data services on GCP Google Kubernetes**: Example walkthroughs for the creation of a GKE cluster and the deployment of Azure Arc data services on that cluster.
- **Data services on upstream Kubernetes** (Kubeadm): Example walkthroughs for the creation of a single-node Kubernetes cluster and the deployment of Azure Arc data services on that cluster.

The GitHub repository provides excellent jumpstart documentation. It covers a wide variety of scenarios, and they are always looking for more if you are interested in contributing.

We have covered a lot of material in our overview of Azure Arc. Hopefully, these sections have provided you with a solid starting point as you begin your proof-of-concept projects. We must now turn our attention to Azure Stack, another critical piece of the Azure hybrid and multicloud strategy.

Modernizing your datacenter with Azure Stack

The Azure Stack family of solutions can extend Azure services and capabilities to all of your locations, such as your on-premises datacenter, remote offices, and edge devices. You can also create and run hybrid applications consistently across location and environment boundaries, meeting the requirements of diverse workloads.

Azure Stack is a family of three products:

- **Azure Stack Hub**: A cloud-native integrated system that brings Azure cloud services on-premises
- **Azure Stack HCI**: An HCI that modernizes datacenters by refreshing virtualization hosts using modern software-defined storage and networking combined with Hyper-V for compute
- **Azure Stack Edge**: A cloud-managed edge appliance for running AI/ML, IoT solutions, and edge computing workloads

These three Azure Stack solutions can be positioned as follows:

1. Azure Stack Hub:
   - Private, autonomous cloud—enabling both completely disconnected scenarios as well as connected-hybrid-cloud scenarios, while maintaining operational consistency with Azure
   - App modernization (cloud-native apps)
   - Data sovereignty, regulatory, and compliance scenarios
   - Connected and disconnected or airgapped scenarios

2. Azure Stack HCI:
   - Hybrid and hyperconverged solution integrated with Azure
   - Modernization of on-premises architecture eliminating the complexity of SAN storage
   - Scalable virtualization, storage, and networking
   - High-performance workloads
   - Remote and branch offices

3. Azure Stack Edge:
   - Azure-managed appliance as a service
   - Hardware-accelerated compute, AI/ML, and IoT applications
   - Low-latency workloads
   - Cloud storage gateway capabilities

This chapter is focused on Azure Stack Hub and Azure Stack HCI. Further information on Azure Stack Edge can be found here: https://azure.microsoft.com/products/azure-stack/edge/.

Now we have briefly reviewed the three Azure Stack solutions, let's discuss one of its varieties, Azure Stack Hub, in detail.

Azure Stack Hub overview

Azure Stack Hub brings Azure services on-premises, in your datacenter, and it is an extension of Azure. Using the same cloud platform across its digital estate, your organization can confidently let business requirements drive technology decisions instead of having technology limitations impact business decisions.

Why use Azure Stack Hub?

While native Azure offers developers comprehensive services for building modern apps, latency, sporadic connectivity, data regulations, and compliance, it can pose challenges for cloud-based applications. To solve these issues, Azure and Azure Stack Hub opens possibilities for new hybrid cloud scenarios for applications, whether intended for customers or internal use:

- **On-premises cloud apps**: Using Azure services and container-based microservices and serverless architectures, you can extend current apps or create new ones to leverage cloud-style advantages and maintain operations consistency. This consistency across native-cloud Azure and an on-premises Azure Stack Hub allows one set of DevOps processes, helping accelerate app modernization and build robust mission-critical apps.

- **Cloud app compliance**: Azure Stack Hub lets you combine on-premises requirements with cloud benefits. You can create and develop apps in an Azure environment for deployment on-premises with Azure Stack Hub with no code changes to satisfy regulatory and compliance requirements. Examples include apps for expense reporting, financial reporting, foreign exchange trading, and global audit.

- **Intermittent or no connectivity**: You can use Azure Stack Hub with only intermittent or even no connectivity to Azure or the internet. Remote production sites, ships, and military applications are examples. Data can be processed in your Azure Stack Hub installation, then aggregated in Azure at a convenient time for additional analytics, avoiding issues of latency or permanent connection.

Use cases for Azure Stack Hub

Among the many possibilities, we list selected examples here from six major sectors:

1. Financial services:
   - Modernize mission-critical apps with a microservices architecture supported by Azure services and containers.
   - Satisfy regulatory requirements while streamlining operations by running cloud apps on-premises and keeping data, applications, and identities secure.
   - Leverage real-time insights, mitigate risks while avoiding latency, and enhance the customer experience through AI and other apps that run in local HCI.

2. Government:
   - Use predictive maintenance on vehicle fleets and manage buildings for energy efficiency via IoT solutions that incorporate machine learning.
   - Enhance services for citizens through better on-premises application and database performance, including for legacy applications.
   - Remain compliant and enhance governance by using a consistent set of tools for application and data management.

3. Manufacturing:
   - Use hybrid cloud capabilities to improve productivity and efficiency by running Azure services without the need for a permanent connection to the internet.
   - Improve employee safety through AI applications at the edge to alert on hazards, and avoid machine breakdown through predictive maintenance.
   - Monitor output from production stages in real time to enhance quality and reduce defects and damage.

4. Retail:
   - Boost customer satisfaction through smarter use of data, including the analysis of promotions and customer interest local to each retail outlet.
   - Optimize the availability of products by monitoring inventory and rate of purchase to re-order the right quantities at the right time.
   - Reduce losses due to shrinkage, shoplifting, return fraud, or other inventory impacts through real-time intelligence from video and application data.

5. Energy:
    - Use essential cloud services in activities and areas that are not connected to the internet, such as remote exploration sites and power grids.
    - Reduce and avoid expensive equipment failures, resolve remote site issues faster, and find and fix problems before they impact worker safety.
    - Store and process data locally to immediately optimize output and operations for oil wells, refineries, power stations, wind farms, and more.

6. Health:
    - Modernize legacy systems and help protect patients from system malfunctions, while integrating medical devices and apps in cloud-like and hyperconverged environments.
    - Improve clinical environments and resource utilization, including operating room performance and ward occupancy, by moving infrastructure to flexible cloud configurations.
    - Optimize data analytics and health records management through local aggregation, processing, and medical records storage in a scalable, containerized infrastructure.

App-, PaaS-, and IaaS-level uses for Azure Stack Hub

Azure Stack Hub is relevant at each of these three conventional cloud resource levels:

- **App level**: Azure Stack Hub can support your app deployment and operation with modern DevOps practices. DevOps teams can maximize productivity and results by using infrastructure as code, **continuous integration/continuous deployment (CI/CD)**, Azure-consistent VM extensions, and other features.

- **Platform as a Service (PaaS) level**: Azure Stack Hub is also a platform for building and running apps that require on-premises PaaS services such as Event Hubs and Web apps. The services are available in Azure Stack Hub in the same way as in Azure, for a unified hybrid development and runtime environment.

- **Infrastructure as a Service (IaaS) level**: Azure Hub Stack enables strongly isolated, self-service resources with detailed tracking of usage, and multi-tenant usage reporting. This makes it an ideal IaaS solution for enterprise private clouds and service providers. Azure Stack Hub, at its core, is an IaaS platform, and we've explored the https://azure.microsoft.com/blog/azure-stack-iaas-part-one/ series.

Azure Stack Hub architecture

Azure Stack Hub integrated systems are packaged as groups of 4-16 servers (called **scale units**) delivered to your datacenter after being built by trusted hardware partners. The integrated system with its hardware and software provides a solution that offers cloud-paced innovation and simplified IT management. Azure Stack Hub uses industry-standard hardware and enables the same management tools as for Azure subscriptions, so you can use consistent DevOps processes independently of any connection to Azure. Azure Stack Hub integrated systems are supported both by Microsoft and the hardware partner.

Azure Stack Hub identity provider

Azure Stack Hub uses one of the following two identity providers: Azure AD or **Active Directory Federation Services (AD FS)**. Many internet-connected, hybrid configurations use Azure AD. On the other hand, disconnected deployments require the use of AD FS. Azure Stack Hub resource providers, along with other apps, function similarly with Azure AD or AD FS. Note that Azure Stack Hub has its own instance of Active Directory and an Active Directory Graph API.

How is Azure Stack Hub managed?

In the same way that Microsoft delivers Azure services to tenants, you can use Azure Stack Hub operations to provide different services and apps to your tenant users. This is because Azure and Azure Stack Hub use the same operations model.

Azure Stack Hub introduces a new role called **Operator**. This is an administrator-level function that is used to manage, monitor, and configure Azure Stack Hub. It is a critical role for the Azure Stack Hub environment, which requires a wide range of skills and knowledge—which are reflected in the Microsoft Certified: Azure Stack Hub Operator Associate (https://docs.microsoft.com/learn/certifications/azure-stack-hub-operator?WT.mc_id=Azure_blog-wwl) certification and Exam AZ-600: Configuring and Operating a Hybrid Cloud with Microsoft Azure Stack Hub (https://docs.microsoft.com/learn/certifications/exams/az-600?WT.mc_id=Azure_blog-wwl).

To prepare for this exam, we have published a set of materials (including the Azure Stack Hub Foundation Core https://github.com/Azure-Samples/Azure-Stack-Hub-Foundation-Core/tree/master/ASF-Training materials) listed in TechCommunity blogs: https://techcommunity.microsoft.com/t5/azure-stack-blog/azure-stack-hub-operator-certification-az-600/ba-p/2024434.

Azure Stack Hub can be managed using three options:

- Administrator portal
- User portal
- PowerShell

The administrator portal can be used for management actions on Azure Stack Hub such as status monitoring or health maintenance of the integrated system, adding marketplace items, adding capacity, and adding new resource providers to enable new PaaS services. The Azure Stack Hub administration portal quickstart (https:// docs.microsoft.com/azure-stack/operator/azure-stack-manage-portals) has more information on using the administrator portal to manage Azure Stack Hub.

Azure Stack Hub Quickstart templates are also available with examples for the deployment of resources, from a simple VM installation to more complex deployments such as Exchange and SharePoint.

As an Azure Stack Hub Operator, you can enable various resource types for your users. For example, this can include SQL and MySQL servers, as well as VM custom images, app services, Azure functions, event hubs, and others. As an Operator, you can also manage capacity issues, create usage offers and subscriptions for tenants, and respond to alerts.

Users can leverage the self-service capabilities of the user portal to consume cloud resources such as web apps, storage accounts, and VMs. Users consume services that are made available by the Operator, and they can provision, monitor, and manage services that they have subscribed to, such as storage, web apps, and VMs. Users have the choice of managing their environment with either the user portal or PowerShell.

Azure Stack Hub provides a multi-tenant environment. This has enabled various service providers (including cloud solution providers, managed service providers, and **independent software vendors (ISVs)**) to build and offer value on top of the Azure Stack Hub platform and deliver this to multiple customers, each customer being isolated and secured in their own respective Azure Stack Hub subscriptions.

Resource providers

Resource providers are services that form the foundation for all Azure Stack Hub IaaS and PaaS services. There are three **foundational IaaS resource providers** in Azure Stack Hub:

- **Compute resource provider**: This allows the creation of VMs by your Azure Stack Hub tenants. With this provider, both VMs and VM extensions can be created.

- **Network resource provider**: Enables the creation of network security groups, public IPs, virtual networks, and software load balancers.

- **Storage resource provider**: Supports the creation of Azure Blob, Table, and Queue Storage services. Azure Key Vault (used to create and manage secret keys) is also supported through this resource provider.

You can also deploy and use any of these **optional PaaS resource providers** with Azure Stack Hub:

- **App Service**: A PaaS offering that allows customers to create web, API, and Azure Functions apps for any device or platform. Both your internal and external customers can automate their business processes and integrate your apps with their on-premises apps. These customer apps can be run by Azure Stack Hub cloud operators on shared or dedicated VMs that are fully managed.

- **Event Hubs**: Event Hubs on Azure Stack Hub allows you to realize hybrid cloud scenarios. Streaming- and event-based solutions are supported, for both on-premises and Azure cloud processing. Whether your scenario is hybrid (connected), or disconnected, your solution can support the processing of events/streams at a large scale. Your scenario is bound only by cluster size, which you can provision according to your needs. (Source: https://docs. microsoft.com/azure/event-hubs/event-hubs-about)

- **IoT Hub (preview)**: IoT Hub on Azure Stack Hub allows you to create hybrid IoT solutions. IoT Hub is a managed service, acting as a central message hub for bi-directional communication between your IoT application and the devices it manages. You can use IoT Hub on Azure Stack Hub to build IoT solutions with reliable and secure communications between IoT devices and an on-premises solution backend. (Source: https://github.com/MicrosoftDocs/azure-docs/blob/master/articles/iot-hub/about-iot-hub.md)

- **SQL Server**: This allows you to provide SQL databases as a service to your Azure Stack Hub tenants by providing a connector to an SQL Server instance.

- **MySQL Server**: This lets you make MySQL databases available as an Azure Stack Hub service by providing a connector to a MySQL server instance.

Azure Stack Hub administration basics

There are several fundamental aspects of Azure Stack Hub administration that you must know about, including understanding the builds, the choice of management tools, the typical responsibilities of an Azure Stack Hub Operator, and what to communicate to your users to help them become more productive.

Understanding the builds

The following are two build-related components you should be aware of:

- **Integrated systems**: Update packages distribute updated versions of Azure Stack Hub for Azure Stack Hub integrated systems. You can import these packages and apply them by using the **Updates** tile in the administrator portal.
- **Development kit**: The **Azure Stack Development Kit (ASDK)** is available as a sandbox for you to evaluate Azure Stack Hub and build and test apps in a non-production environment. The ASDK deployment page provides more information on deployment: https://docs.microsoft.com/azure-stack/asdk/asdk-install.

Innovation in Azure Stack Hub is rapid, with regular releases of new builds. If you are running ASDK and you want to move to the latest build and the most recent features of Azure Stack Hub, you cannot simply apply update packages—you must redeploy the ASDK. The ASDK documentation on the Azure website reflects the latest release build: https://docs.microsoft.com/azure-stack/asdk/.

Staying aware of available services

Azure Stack Hub supports an evolving subset of Azure services. You will need to be aware of the services that you can offer your Azure Stack Hub users at any given time.

Foundational services

By default, Azure Stack Hub includes the following foundational services at deployment:

- Compute
- Storage
- Networking
- Key Vault

These foundational services allow you to propose IaaS to your users with minimal configuration.

Additional services

Following are the currently supported additional PaaS services:

- App Service
- Azure Functions
- SQL and MySQL RPs
- Event Hubs
- IoT Hub (in preview)
- Kubernetes (in preview)

Before you can offer these additional services to your users, further configuration is needed. Check out the *Tutorials* and the *How-to guides* sections here for more information: https://docs.microsoft.com/azure-stack/operator/.

What management tools can you use?

The administrator portal is the easiest way to learn basic management concepts. You can also use PowerShell for Azure Stack Hub management, although there are certain preparation steps. It may help to familiarize yourself with the way PowerShell applies to Azure Stack Hub, by visiting *Get started with PowerShell on Azure Stack Hub*: https://docs.microsoft.com/azure-stack/user/azure-stack-powershell-overview.

As an underlying deployment, management, and organization mechanism, Azure Stack Hub uses Azure Resource Manager. To find out more about Resource Manager so that you can manage Azure Stack Hub and assist in supporting users, visit *Getting Started with Azure Resource Manager whitepaper*: https://download.microsoft.com/download/E/A/4/EA4017B5-F2ED-449A-897E-BD92E42479CE/Getting_Started_With_Azure_Resource_Manager_white_paper_EN_US.pdf.

Typical administrator/provider responsibilities

Users see the Azure Stack Hub Operator role as making available to them the services that they require. Accordingly, decide which services will be proposed, then create plans, offers, and quotas to make those services available. See also *Overview of offering services in Azure Stack Hub*: https://docs.microsoft.com/azure-stack/operator/service-plan-offer-subscription-overview.

You will also need to add items to Azure Stack Hub Marketplace. Downloading marketplace items to Azure Stack Hub from Azure is the easiest way to do this.

> **Note**
>
> You should use the user portal rather than the administrator portal to test the user availability of your services and the plans and offers for these services.

In addition to providing services, there are periodic operator tasks to perform to ensure Azure Stack Hub continues to run correctly. These tasks include:

- The creation of user accounts for Azure AD deployment (https://docs.microsoft.com/azure-stack/operator/azure-stack-add-new-user-aad) or AD FS deployment (https://docs.microsoft.com/azure-stack/operator/azure-stack-add-users-adfs).
- RBAC role assignment (not restricted to administrators) (https://docs.microsoft.com/azure-stack/operator/azure-stack-manage-permissions).
- Infrastructure health monitoring (https://docs.microsoft.com/azure-stack/operator/azure-stack-monitor-health).
- Network management (https://docs.microsoft.com/azure-stack/operator/azure-stack-viewing-public-ip-address-consumption) and storage resource management (https://docs.microsoft.com/azure-stack/operator/azure-stack-manage-storage-accounts).
- Replacing bad hardware, such as a failed disk (https://docs.microsoft.com/azure-stack/operator/azure-stack-replace-disk).

Communicating with your users

Your users will need to know how to connect to the Azure Stack Hub environment, how to subscribe to offers, and how to use Azure Stack Hub services. The Azure Stack Hub user documentation (https://docs.microsoft.com/azure-stack/user/) is a useful information source.

Understanding how to use Azure Stack Hub services

Before users work with services and build apps in Azure Stack Hub, there are prerequisites such as specific versions of PowerShell and APIs. There are also some feature differences between an Azure service and the corresponding Azure Stack Hub service. Users should therefore familiarize themselves with the following content:

- Key considerations for using services or building apps for Azure Stack Hub: https://docs.microsoft.com/azure-stack/user/azure-stack-considerations
- Considerations for VMs in Azure Stack Hub: https://docs.microsoft.com/azure-stack/user/azure-stack-vm-considerations
- Differences and considerations for Azure Stack Hub storage: https://docs.microsoft.com/azure-stack/user/azure-stack-acs-differences

The information in these three articles gives an overview of the differences between services in Azure and Azure Stack Hub. It is additional information to the global Azure documentation for a given Azure service.

Connections for users to Azure Stack Hub

In the ASDK case, users can configure a **virtual private network (VPN)** connection to connect to the ASDK environment if they are not using Remote Desktop. More information is available here: https://docs.microsoft.com/azure-stack/asdk/asdk-connect.

Users will also want to know how to connect via the user portal (https://docs.microsoft.com/azure-stack/user/azure-stack-use-portal) or by using PowerShell. The user portal URL will vary per deployment in an integrated systems environment, so be sure to provide your users with the correct address.

If the portal is published on the internet, users could also use things such as Cloud Shell to manage and create resources on Azure Stack Hub (please note that the Cloud Shell service itself is not available on Azure Stack Hub—you would need to be in a connected environment).

Before using services via PowerShell, users may need to register resource providers, such as the networking resource provider that manages resources like load balancers, network interfaces, and virtual networks. Users must install PowerShell (https://docs.microsoft.com/azure-stack/operator/powershell-install-az-module), download additional modules (https://docs.microsoft.com/azure-stack/operator/azure-stack-powershell-download), and configure PowerShell (which includes resource provider registration): https://docs.microsoft.com/azure-stack/user/azure-stack-powershell-configure-user.

Subscribing to an offer

Users must also subscribe to an offer (https://docs.microsoft.com/azure-stack/operator/azure-stack-subscribe-plan-provision-vm) that an operator has created before they can use services. Offers are groups of one or more plans that enable Azure Stack Hub Operators to control things such as trial offers and capacity planning, and even delegate the creation of Azure Stack Hub subscriptions to other users.

Getting support

Azure Stack Hub covers many areas. Before contacting Microsoft Support, review this web page. It provides common questions and issues with troubleshooting links: https://docs.microsoft.com/azure-stack/operator/azure-stack-servicing-policy.

In addition to the issues outlined there, the following are some additional support considerations.

Integrated systems

An integrated system benefits from a coordinated escalation and resolution process between Microsoft and Microsoft's **original equipment manufacturer (OEM)** hardware partners.

Microsoft Support can help if there is a cloud services issue. To open a support request, select the help and support icon (question mark) in the upper-right corner of the administrator portal. Then select **Help + support** and then **New support request** under the **Support** section.

Your OEM hardware vendor is your first point of contact if there is an issue with deployment, patch, update, hardware (including field replaceable units), or any hardware-branded software, such as software running on the hardware lifecycle host.

You should contact Microsoft Support for help with any other problems.

Azure Stack Development Kit

ASDK is a single-node deployment of Azure Stack Hub that you can download and use for free. All ASDK components are installed in VMs running on a single-host computer that must have enough resources. ASDK is meant to provide an environment in which you can evaluate Azure Stack Hub and develop modern apps using APIs and tooling consistent with Azure in a non-production environment.

As ASDK is a free product, there is no formal support offered. You can ask support-related questions about ASDK in the regularly monitored Microsoft forums (https://social.msdn.microsoft.com/Forums/azure/home?forum=azurestack). To access the forums, select the help and support icon (question mark) in the upper-right corner of the administrator portal, then select **Help + support**, and then select **MSDN Forums** under the **Support** section. However, there is no official support offered through Microsoft Support as the ASDK is an evaluation environment.

Azure Stack Hub update management

You can help to keep Azure Stack Hub up to date by applying full and express updates, hotfixes, and driver and firmware updates from the OEM. Remember, however, that Azure Stack Hub update packages are intended for integrated systems and cannot be applied to ASDK, for which redeployment is required. See *Redeploy* ASDK at https://docs.microsoft.com/azure-stack/asdk/asdk-redeploy.

Update package types

For integrated systems, there are three different types of update packages:

- **Azure Stack Hub software updates**: These updates are downloaded directly from Microsoft and can include Azure Stack Hub feature updates, Windows Server security, and non-security-related updates. Each update package can either be of the type **full** or **express**:
  - Full update packages update the physical host operating systems in the scale unit. They require a larger maintenance window.
  - Express update packages are scoped and don't update the underlying physical host operating systems.
- **Azure Stack Hub hotfixes**: These time-sensitive updates address a specific issue. Each hotfix is released with a corresponding Microsoft Knowledge Base article that details the issue, cause, and resolution. Hotfixes are downloaded and installed like the full update packages for Azure Stack Hub. You can read more about them here: https://docs.microsoft.com/azure-stack/operator/azure-stack-servicing-policy#hotfixes.
- **OEM hardware vendor–provided updates**: These updates are managed by each hardware vendor. They typically contain driver and firmware updates, and each vendor hosts these on its own site.

When to update

The updates above are released at different frequencies:

- **Azure Stack Hub software updates**: Multiple full and express software update packages a year are released by Microsoft.
- **Azure Stack Hub hotfixes**: These are time-sensitive and can be released at any time. When you upgrade from one major version to another, if there are any hotfixes in the new version, they are automatically installed.
- **OEM hardware vendor-provided updates**: These updates are released by OEM hardware vendors on an as-needed basis.

You must keep your Azure Stack Hub environment on a supported Azure Stack Hub software version if you want to continue to receive support. For more information, visit this link to read the Azure Stack Hub Servicing Policy: https://docs.microsoft.com/azure-stack/operator/azure-stack-servicing-policy.

Checking available updates

The notification of updates depends on factors such as your internet connectivity and the type of update:

- For software updates and hotfixes released by Microsoft, you will see an update alert appear in the **Update** pane for Azure Stack Hub instances that are connected to the internet. Restart the infrastructure management controller VM if this blade is not displayed. For instances that are not connected, you can subscribe to the RSS feed (https://azurestackhubdocs.azurewebsites.net/xml/hotfixes.rss) to receive notification of each hotfix release.
- Notifications about updates provided by OEM hardware vendors depend on your communications with your manufacturer. For more information, visit the Apply Azure Stack Hub OEM updates page: https://docs.microsoft.com/azure-stack/operator/azure-stack-update-oem.

Updating from one major version to the next

Updating from one major version to the next major version must be done in the correct order and step by step. You cannot skip a major version update. For more details, please review this link: https://docs.microsoft.com/azure-stack/operator/azure-stack-updates#how-to-know-an-update-is-available.

Hotfixes within major versions

Within a major version release, there can be multiple hotfixes. These are cumulative, with the last hotfix package for that major version including all previous hotfixes for it. For more information, see https://docs.microsoft.com/azure-stack/operator/azure-stack-servicing-policy#hotfixes.

The update resource provider

Azure Stack Hub handles the application of Microsoft software updates through an update resource provider. The provider verifies that updates have been applied across the Service Fabric apps and their runtimes, all of the physical hosts, and all VMs and the services associated with them.

During the update process, you have access to a high-level status overview in Azure Stack Hub, where you can see progress for the different subsystems.

We have covered the basics of Azure Stack Hub. Next, we move on to Azure Stack HCI, an HCI cluster solution.

Azure Stack HCI solution overview

Azure Stack HCI is designed for hosting virtualized Linux and Windows workloads on validated partner hardware. Designed for the hybrid environment (on-premises and cloud), Azure Stack HCI combined with Azure hybrid services provides capabilities such as disaster recovery and backup options for your VMs, as well as cloud-based monitoring. You also get a single pane of glass for all your deployments and the ability to manage the cluster with tools such as Windows Admin Center, System Center, and PowerShell. In addition, AKS on Azure Stack HCI is available. It is an on-premises implementation of the AKS orchestrator—see https://docs.microsoft.com/azure-stack/aks-hci/overview.

Azure Stack HCI software is available for download here: https://azure.microsoft.com/products/azure-stack/hci/hci-download/. An on-premises solution with built-in hybrid-cloud connections, it is billed to your Azure subscription as an Azure service.

Azure Stack HCI use cases

There are a number of use cases for Azure Stack HCI, which we will discuss now.

Datacenter consolidation and modernization

Using Azure Stack HCI to refresh older virtualization hosts with possible consolidation has several potential benefits. It can enhance scalability and facilitate management and security for your environment. It can also reduce the footprint and total cost of ownership by replacing legacy SAN storage. Unified tools and interfaces, together with a single point of support, simplify systems administration and operations.

Hybrid infrastructure management for branch offices

Management of hybrid infrastructure can be challenging for enterprises with branch offices. When multiple locations must function without dedicated IT staff, it becomes more complicated to keep identity services synchronized, make data backups, and deploy applications. In this situation, businesses often need weeks or months to roll out application updates across multiple offices and infrastructures. Better solutions are needed to deploy application and identity changes quickly and easily across remote offices while enabling centralized monitoring for anomalies and violations.

Consider the example of an international bank with 300 offices worldwide, for which each global update of all offices requires a year. To compound the problem, the multiple locations make it difficult to consistently correct configurations and avoid potential security risks such as open ports. New and updated app rollouts to branch offices are also a problem for any company that has double- or triple-digit numbers of such sites. This is especially true when remote sites need to run apps on local servers for reasons of latency or restricted internet connectivity.

To help you overcome these challenges, Azure Stack HCI offers HCI based on industry-standard x86 servers with software-defined compute, storage, and networking. Using Azure Arc integration built into the Windows Admin Center, you can easily begin using the Azure portal cloud to manage your HCI. Reducing or removing the need for local IT staff, you can meet the evolving IT demands of branch offices, retail stores, and field locations, deploying container-based apps anywhere and essential business applications in highly available VMs. Azure Monitor then allows you to view system health across these different domains.

Remote and branch offices

Azure Stack HCI provides an affordable way to modernize remote and branch offices, including retail stores and field sites (two-server cluster solutions are currently available for less than $20,000 per location). Nested resiliency means that volumes can remain online and accessible even in the event of multiple simultaneous hardware failures. Cloud witness technology enables the use of Azure as the lightweight tiebreaker for cluster quorum. This prevents split-brain conditions yet does not involve the cost of a third host. In the Azure portal, you are given a centralized view of remote Azure Stack HCI deployments.

Virtual desktops

Azure Stack HCI provides like-local performance for your on-premises virtual desktops. This is ideal if you need to support data sovereignty for your users as well as low latency.

Azure integration benefits

Azure Stack HCI lets you leverage combined cloud and on-premises resources via hybrid infrastructure, with native cloud monitoring, security, and backup.

First, your Azure Stack HCI cluster must be registered with Azure. Then, you will be able to utilize the Azure portal for:

- **Monitoring**: You are given an overall view of your Azure Stack HCI clusters. Here, you can group them by resource group and tag them. New upcoming features will also enable the creation and management of VMs from the portal.
- **Billing**: Use your Azure subscription to pay for Azure Stack HCI (after the free preview phase).
- **Support**: Azure Stack HCI support is available through a Standard or Professional Direct Azure support plan.

These additional Azure hybrid services are also available for subscription:

- Azure Site Recovery for **disaster recovery as a service (DRaaS)** and high availability.
- Azure Monitor, for a centralized view of events on your apps, infrastructure, and network, including AI-driven advanced analytics.
- Cloud Witness, to leverage Azure as the lightweight cluster quorum tiebreaker.
- Azure Backup to protect data by storing it offsite, also achieving protection against ransomware.
- Azure Update Management to assess and deploy updates for Azure-hosted and on-premises Windows VMs.
- Azure Network Adapter to use a Point-to-Site VPN for connections between your Azure VMs and your on-premises resources.
- Azure File Sync for cloud syncing of your file server.

See also *Connecting Windows Server to Azure hybrid services*: https://docs.microsoft. com/windows-server/manage/windows-admin-center/azure/index.

Azure Stack HCI prerequisites

Initially, you will require:

- A cluster of at least two servers chosen from the Azure Stack HCI catalog and your elected Microsoft hardware partner.
- A subscription to Azure.
- An internet connection for HTTPS outbound transmissions every 30 days or more frequently for each server in the cluster.

If your clusters span several sites, you will require at minimum a 1 Gb connection between sites (but preferably a 25 Gb RDMA connection), with a 5 ms round trip average latency if you plan to have synchronous replication with writes being done simultaneously at both sites.

If you aim to use **software-defined networking (SDN)**, plan also for the creation of Network Controller VMs on a **virtual hard disk (VHD)** for the Azure Stack HCI operating system (for more information, visit the *Plan to deploy Network Controller* page: https://docs.microsoft.com/azure-stack/hci/concepts/network-controller).

See also:

- System requirements: https://docs.microsoft.com/azure-stack/hci/concepts/system-requirements
- AKS requirements on Azure Stack HCI (for AKS on Azure Stack HCI): https://docs.microsoft.com/azure-stack/aks-hci/overview#what-you-need-to-get-started

Hardware partners

By ordering validated Azure Stack HCI configurations from your elected Microsoft partner, you can start operations without undue design or deployment time. Implementation and support are also available via Microsoft partners from a single point of contact through joint support agreements. Options include buying validated nodes, or an integrated system with pre-installation of the Azure Stack HCI operating system together with driver and firmware update partner extensions.

Deployment options

- **Integrated systems**: Buy validated servers with a pre-installation of Azure Stack HCI from a hardware partner.

- **Validated nodes**: Buy validated bare metal servers from a hardware partner, then sign up for the Azure Stack HCI service. Go to the Azure portal, where you can then obtain the Azure Stack HCI operating system download.

- **Repurposed hardware**: Repurpose your existing hardware. For details on this path, review the following link: https://docs.microsoft.com/azure-stack/hci/deploy/migrate-cluster-same-hardware.

You can read more about Azure Stack HCI here: https://azure.microsoft.com/overview/azure-stack/hci.

You can also check out the Azure Stack HCI catalog. It has details on more than 100 solutions from Microsoft partners. You can find the online catalog here: https://azure.microsoft.com/products/azure-stack/hci/catalog/.

Software partners

Microsoft partners are developing software to build on the Azure Stack HCI platform without needing to replace the tools already familiar to IT administrators. To review a list of highlighted ISVs and their applications, check out this link: https://docs.microsoft.com/azure-stack/hci/concepts/utility-applications.

Licensing, billing, and pricing

Azure Stack HCI billing uses the number of physical processor cores to calculate the monthly subscription fee, instead of a perpetual license. The information on the number of cores that you are using is uploaded and evaluated for billing automatically when you connect to Azure. In this way and because more VMs does not mean higher costs, denser virtual environments lead to more attractive pricing.

Management tools

You have full admin cluster rights with Azure Stack HCI, letting you directly manage:

- **Failover clustering**: https://docs.microsoft.com/windows-server/failover-clustering/failover-clustering-overview

- **Hyper-V on Windows Server**: https://docs.microsoft.com/windows-server/virtualization/hyper-v/hyper-v-on-windows-server

- **SDN**: https://docs.microsoft.com/windows-server/networking/sdn/

- **Storage Spaces Direct**: https://docs.microsoft.com/windows-server/storage/storage-spaces/storage-spaces-direct-overview

You also have these management tools available for use:

- **PowerShell**: https://docs.microsoft.com/powershell/
- **Server Manager**: https://docs.microsoft.com/windows-server/administration/server-manager/server-manager
- **System Center**: https://www.microsoft.com/system-center
- **Windows Admin Center**: https://docs.microsoft.com/windows-server/manage/windows-admin-center/overview
- Other non-Microsoft tools such as 5Nine Manager

Getting started with Azure Stack HCI and Windows Admin Center

These sections apply for Azure Stack HCI, version 20H2. They assume that you have already set up a cluster in your Azure Stack HCI installation and they give instructions for connecting to the cluster and monitoring cluster and storage performance.

Installing Windows Admin Center

Windows Admin Center is a locally deployed, browser-based app for managing Azure Stack HCI. Windows Admin Center can be installed on a server in **service mode**, but it is simpler to install it on a local management PC in **desktop mode**. In service mode, tasks that require CredSSP, such as cluster creation and installing updates and extensions, necessitate the use of an account that is a member of the Gateway Administrators group on the Windows Admin Center server. For further details, check out https://docs.microsoft.com/windows-server/manage/windows-admin-center/configure/user-access-control#gateway-access-role-definitions.

Adding and connecting to an Azure Stack HCI cluster

After installing Windows Admin Center, you can add a cluster to be managed from the Windows Admin Center dashboard's main overview page. The dashboard also shows you information on CPU, memory, and storage usage, together with alerts and health information about servers, drives, and volumes. Cluster performance information such as **input/output operations/second** (**IOPS**) and latency by the hour, day, week, month, or year is shown at the bottom of the dashboard displays.

Monitoring individual components

The **Tools** menu to the left of the dashboard allows you to drill down on any component of the cluster to view summaries and inventories of virtual machines, servers, volumes, drives, and virtual switches. Within the dashboard, the Performance Monitor tool lets you view, compare, and add performance counters for Windows, apps, or devices in real time.

Viewing the cluster dashboard

To view cluster dashboard information on cluster health and performance, select the cluster name under **All connections**. Then, on the left under **Tools**, select **Dashboard**. You can then see:

- Average cluster latency in milliseconds
- Cluster event alerts
- A list of:
  - Disk drives available on the cluster
  - Servers joined to the cluster
  - VMs running on the cluster
  - Volumes available on the cluster
- Total cluster:
  - CPU usage for the cluster
  - IOPS
  - Memory usage for the cluster
  - Storage usage for the cluster

For more information, visit this link: https://docs.microsoft.com/azure-stack/hci/manage/cluster.

Using Azure Monitor for monitoring and alerts

You can also use Azure Monitor to collect events and performance counters for analysis and reporting, react to specific conditions, and receive email notifications. By clicking Azure Monitor from the **Tools** menu, you will make a direct connection from Windows Admin Center to Azure.

Collecting diagnostics information

The **Diagnostics** tab in the **Tools** menu lets you collect troubleshooting information for your cluster. This information may also be requested by Microsoft Support if you ask for assistance.

Managing VMs with Windows Admin Center

You can use Windows Admin Center to spin up and manage your VMs on Azure Stack HCI (Azure Stack HCI, version 20H2 and Windows Server 2019). Functionality includes:

- Creating a new VM.
- Listing VMs on a server or in a cluster.
- Viewing VM details, including detailed information and performance charts from the dedicated page for a given VM.
- Viewing aggregate VM metrics to see overall resource usage and performance for all VMs in a cluster.
- Changing VM settings for memory, processors, disk size, and more (note that you will need to stop a VM first before changing certain settings).

Additional functionality includes:

- Moving a VM to another server or cluster
- Joining a VM to a domain
- Cloning a VM
- Importing and exporting a VM
- Viewing VM export logs
- Protecting VMs with Azure Site Recovery (an optional value-add service)

Note that you can also connect to a VM in other ways than just through Windows Admin Center:

- Via a Hyper-V host using a **Remote Desktop Protocol** (**RDP**) connection
- By using Windows PowerShell

You can also make Storage Spaces Direct changes for your cluster and change general cluster settings. These can include:

- Cluster witnesses
- Node shutdown behavior
- Setting and managing access points
- Traffic encryption
- VM load balancing

You can also monitor your Azure HCI cluster via Azure Monitor.

Comparing Azure Stack Hub and Azure Stack HCI

Azure Stack HCI and Azure Stack Hub can both play an essential role in your hybrid and multicloud strategy. Here is a quick summary of their differences:

- **Azure Stack Hub**: This allows you to run cloud apps on-premises, when you are disconnected, or to meet regulatory requirements, by leveraging consistent Azure services.
- **Azure Stack HCI**: This allows you to run virtualized apps on-premises, strengthen and replace aging server infrastructure, and connect to Azure for cloud services.

Azure Stack Hub offers innovative processes, though may require new skills. It brings Azure services into your datacenter. On the other hand, with Azure Stack HCI, you can use existing skills and familiar processes, and you connect your datacenter to Azure services. It may also be useful to compare what Azure Stack Hub and Azure Stack HCI are not designed to do, or possible constraints.

Azure Stack Hub limitations:

- Azure Stack Hub must have a minimum of 4 nodes, as well as its own network switches.
- Azure Stack Hub limits Hyper-V's configurability and feature set in order to remain consistent with Azure.
- Azure Stack Hub does not give you access to underlying infrastructural technologies.

Azure Stack HCI limitations:

- Azure Stack HCI does not natively offer or enforce multi-tenancy.
- Azure Stack HCI does not offer on-premises PaaS features.
- Azure Stack HCI does not include native DevOps toolsets.
- Azure Stack HCI can only work virtualized workloads—no application can be deployed on the metal.

For additional reading, visit https://docs.microsoft.com/azure-stack/operator/compare-azure-azure-stack?view=azs-2008.

Summary

As business environments evolve, software applications may run on different systems on-premises, off-premises, in multiple clouds, and at the edge of networks. Hybrid and multicloud solutions can enable you to successfully deploy and manage your apps if you also create a single, consistent environment across the locations of your organization. In doing so, you offer your developers a unified set of tools for building apps. You can easily move apps and data between locations to ensure efficiency and compliance, and to offer your users an optimal experience.

Azure Arc and the Azure Stack portfolio are two major components of Azure hybrid and multicloud solutions and products. Azure Arc offers unified management to deploy Azure services and extend Azure management anywhere. With Azure Arc enabled servers, you can apply native Azure VM management to Windows and Linux machines outside Azure. With Azure Arc and Kubernetes, you can run specified Azure data services on the infrastructure of your choice.

By comparison, Azure Stack makes Azure services available across cloud and non-cloud locations, letting you create and run hybrid applications for diverse workloads consistently across locations and environments. Azure Stack Hub is a cloud-native integrated system that lets you use Azure cloud services on-premises, while Azure Stack HCI offers HCI that modernizes datacenters and virtualization hosts.

Individually, together, or in combination with other Microsoft and Azure solutions, Azure Arc and Azure Stack can help you meet challenges and make the most of opportunities in sectors ranging over finance, government, manufacturing, retail, energy, health, and more.

You can try out functionality like Azure Arc immediately, using Azure Arc Jumpstart on AKS, AWS Elastic Kubernetes Service, GKE, or in an Azure VM. Some example main steps are as follows:

1. Install the client tools.

2. Create the Azure Arc data controller.

3. Create a managed instance on Azure Arc.

4. Create an Azure Database for the PostgreSQL Hyperscale server group on Azure Arc.

For further information on Azure hybrid and multicloud solutions and products and how to use them successfully for your enterprise or organization, visit these Microsoft web pages:

- Azure Hybrid and multicloud solutions: https://azure.microsoft.com/solutions/hybrid-cloud-app/
- Azure hybrid and multicloud architecture: https://docs.microsoft.com/azure/architecture/browse/?azure_categories=hybrid
- Azure hybrid and multicloud patterns and solutions documentation: https://docs.microsoft.com/hybrid/app-solutions/?view=azs-1910

We hope you have found this chapter helpful in your Azure hybrid and multicloud journey. Next, we will explore how to plan and implement migration to Azure.

Cloud migration: Planning, implementation, and best practices

Customers are accelerating their cloud journey to optimize costs, enhance security and resiliency, and scale on demand. One of the key motivations for organizations to migrate to Azure is to reduce the **total cost of ownership (TCO)** for their IT infrastructure. By migrating to Azure, organizations can move from a **capital expenditure (CapEx)** model with fixed upfront commitments to a more flexible and scalable **operational expenditure (OpEx)** model. With the OpEx model, organizations reduce spending by only paying for resources and services that they consume.

As we will discuss in this chapter, this is only one of many benefits that you will gain by migrating to Azure. First, we'll see that Microsoft provides a framework that will guide you in architecting for reliability, security, and high availability. We will then explore the underlying infrastructure of Azure to help you make the most informed choice to meet your migration goals. After we have built our foundational knowledge in Azure infrastructure, we will discuss some of the common workload migration scenarios and how you can achieve cloud scale and maximize performance.

In this chapter, we will cover the following topics:

- Microsoft's Well-Architected Framework
- Choosing your underlying infrastructure
- Workload migration scenarios
- Achieving cloud scale and maximizing performance
- Enterprise-grade backup and disaster recovery
- Azure migration best practices and support

Let's begin by discussing architectural best practices for planning, designing, and implementing reliable systems in Azure using the Microsoft Well-Architected Framework.

Microsoft Azure Well-Architected Framework

As organizations shifted their workforce to work from home, critical infrastructure and applications were impacted. While some organizations were able to migrate their applications to the cloud, others had to redesign their environment in the cloud. However, designing and deploying a successful workload in the cloud can be challenging, especially when there is a time constraint for an organization to ensure that they continue to support their clients around the world.

When operating in the cloud, there are several design principles and considerations that differ from operating on-premises, such as the way you manage workloads, infrastructure costs, monitoring, security, and performance. Many of the management tasks that we used to do for on-premises applications do not apply to the cloud.

If these considerations for the workloads that we are running in the cloud are not deployed properly, there are various consequences that might impact the cost of the services in Azure and the performance of the application.

To help you and your organization address the complexity of planning, designing, and implementing reliable systems in Azure, Microsoft created the Azure Well-Architected Framework, which provides best practices to build solutions in the Azure cloud. This framework is intended to guide you to improve the quality of your solutions running in Azure and comprises of five main pillars:

- **Cost optimization** so that you can create cost-effective solutions.
- **Operational excellence** to ensure operation continuity by keeping your systems running in production with minimal downtime.
- **Performance efficiency** to have the capacity to scale on-demand and meet business needs when there are peaks in usage.
- **Reliability** to have the ability to rapidly recover from potential failures that can impact the application availability.
- **Security** to protect your applications and data, and quickly respond to potential threats and vulnerabilities.

Based on industry standards, Microsoft is helping customers improve the quality of their workloads. The framework is at the center of the Microsoft Azure Well-Architected initiative. It includes the documentation, reference architectures, and design principles to help you understand and figure out how these workloads can be better designed, implemented, and successfully deployed on the cloud.

Microsoft also included the Azure Well-Architected Review, which is a web application with questions and answers for you to better understand how workloads are being designed and evaluate which points you need to improve.

In addition, Azure Advisor is aligned with the same five pillars of the Azure Well-Architected Framework to provide you with real-time recommendations for all the resources that you are running in the cloud and improve their performance.

We recommend the following Microsoft learning path for the Well-Architected Framework: https://docs.microsoft.com/learn/paths/azure-well-architected-framework/

Many organizations begin their cloud journey on Azure with **Infrastructure as a Service (IaaS)**. In the next section, we will first establish our foundational knowledge in Azure IaaS by learning about the three core services of Azure. Later in this chapter, we will look at other common workload migration scenarios.

Choosing your underlying infrastructure

In this section, we will examine the underlying infrastructure of Azure. At its core, a typical IaaS environment in Azure consists of the following three services:

- Compute
- Networking
- Storage

These three services form the main architectural building blocks in an IaaS environment. We will explore each of these core services in detail, starting with Azure Compute.

Compute

Whether you are deploying new workloads or migrating existing ones to the cloud, Azure Compute has the necessary infrastructure to run your workload.

Here are some of the key capabilities of Azure Compute:

- Azure offers over 700 **virtual machine (VM)** sizes that can address almost every type of workload. From dev/test workloads and mission-critical production workloads to customer-facing applications high-performance computing scenarios.
- You can run your applications on Windows Server and multiple Linux distributions such as Red Hat, SUSE, Ubuntu, CentOS, Debian, Oracle Linux, and CoreOS. Additionally, Microsoft provides 24x7 integrated co-located support with Red Hat and SUSE.
- Azure provides purpose-built bare metal infrastructure to run a growing number of solutions such as SAP HANA, NetApp, or Cray in Azure.
- Azure delivers an **evergreen** foundation with CPUs and GPUs from Intel®, AMD, and NVIDIA, helping you get the best price-to-performance ratio.
- You can choose to run your applications in any available Azure region. You can also take advantage of multiple regions to increase the resiliency of your applications at scale.
- Azure offers a broad range of pricing models, from pay-as-you-go to reservations, which include 1- or 3-year term discounts. This is in addition to optimization tools that offer resizing recommendations, and advice you to get the most out of your Azure environment.

- If you have existing Windows Server subscriptions, SQL Server licenses, or Windows Server licenses with Software Assurance and **Azure Hybrid Benefit**, you can reuse them to save significantly on costs as you move your workloads to Azure. Additionally, through Azure Hybrid Benefit for Linux, you can use your pre-existing on-premises Red Hat and SUSE software subscriptions on Azure.

Perhaps one of the most common migration scenarios for Azure customers is migrating an existing on-premises environment to the cloud with little to no change. This is often referred to as lift and shift, or rehosting. We can leverage virtualization technologies to help migrate existing on-premises workloads to Azure. Without a doubt, virtualization has positively streamlined how organizations are deploying and managing their IT infrastructure on Azure, thanks in part to solutions such as Microsoft Hyper-V and VMware vSphere. Moreover, administrators who are familiar with the characteristics and behaviors of running VMs on-premises can adapt the same skills successfully to Azure.

In the following table, we have provided you with some common use cases and recommended Azure Compute services for each. You can use this table to plan your next cloud migration strategy. We have also included links to each Azure Compute service so that you can learn more about them.

Compute services	Common use cases
Azure App Service: https://azure.microsoft.com/services/app-service/	For developing cloud-native applications for web and mobile with a fully managed platform.
Azure Dedicated Host: https://azure.microsoft.com/services/virtual-machines/dedicated-host/	Deploying Azure VMs on a dedicated physical server to isolate workloads for only your organization to use.
Azure Functions: https://azure.microsoft.com/services/functions/	You can accelerate app development using an event-driven and serverless architecture without any additional setup.
Azure Container Instances: https://azure.microsoft.com/services/container-instances/	To help you easily run containers in Azure with a single command.
Azure Kubernetes Service: https://azure.microsoft.com/services/kubernetes-service/	To simplify the deployment, management, and operation of Kubernetes.

Compute services	Common use cases
Azure Service Fabric: https://azure.microsoft.com/services/service-fabric/	For developing microservices and orchestrating containers on both Windows and Linux.
Azure Batch: https://azure.microsoft.com/services/batch/	To access cloud-scale job scheduling and compute management across 10x, 100x, 1000x VMs.
Azure Cloud Services: https://azure.microsoft.com/services/cloud-services/	If you prefer to use a PaaS technology to deploy web and cloud applications that are scalable, reliable, and inexpensive to operate.
Virtual Machines: https://azure.microsoft.com/services/virtual-machines/	If you need to quickly provision Linux and Windows VMs and maintain complete control over your cloud environment.
Virtual Machine Scale Sets: https://azure.microsoft.com/services/virtual-machine-scale-sets/	To autoscale a large number of load-balanced VMs in minutes to achieve high availability and performance based on demand.
Azure VMware Solution: https://azure.microsoft.com/services/azure-vmware/	If you want to run your VMware workloads natively on Azure.

Table 4.1: Recommended Azure services for common use cases

Now that we have seen the capabilities of Azure Compute, let's look at Azure networking next.

Networking

Azure networking securely connects and delivers your cloud-native and hybrid workloads with low-latency and Zero Trust–based networking services. Zero Trust is a security concept that fosters the idea that organizations shouldn't automatically trust anything inside or outside their perimeters. Instead, anything that tries to connect to its systems must be verified before access is granted.

In the following table, we have provided you with some common use cases and the recommended Azure networking service to choose from. You can use this table to plan your next cloud migration strategy. We have also included links to each Azure networking service so that you can learn more about them:

Network services	Common use cases
Azure Bastion: https://azure.microsoft.com/services/azure-bastion/	For accessing your VMs via private and fully managed RDP and SSH.
Azure Virtual Network: https://azure.microsoft.com/services/virtual-network/	To connect everything from VMs to incoming VPN connections.
Azure ExpressRoute: https://azure.microsoft.com/services/expressroute/	If you need to add private network connectivity from your on-premises network to Azure.
Azure VPN Gateway: https://azure.microsoft.com/services/vpn-gateway/	For using the internet to access Azure Virtual Networks securely.
Azure Virtual WAN: https://azure.microsoft.com/services/virtual-wan/	To securely connect offices, retail locations, and sites with a unified portal.
Azure DDoS Protection: https://azure.microsoft.com/services/ddos-protection/	To protect your applications from DDoS attacks.
Azure Firewall: https://azure.microsoft.com/services/azure-firewall/	You need to add native firewall capabilities with zero maintenance and built-in high availability.
Azure Firewall Manager: https://azure.microsoft.com/services/firewall-manager/	You need a way to manage network security policy and routing centrally.
Azure Load Balancer: https://azure.microsoft.com/services/load-balancer/	Inbound and outbound connections and requests to applications need to be balanced.
Traffic Manager: https://azure.microsoft.com/services/traffic-manager/	You need to route incoming traffic for better performance and availability.
Network Watcher: https://azure.microsoft.com/services/network-watcher/	You need to monitor and diagnose network issues.

Network services	Common use cases
Azure Web Application Firewall: https://azure.microsoft.com/services/web-application-firewall/	You need to use a firewall service for web apps to improve web app security.
Azure Application Gateway: https://azure.microsoft.com/services/application-gateway/	You want to use a web traffic load balancer to manage traffic to your web applications.
Azure DNS: https://azure.microsoft.com/services/dns/	You want to ensure ultra-fast DNS responses and availability to meet your domain requirements.
Azure Private Link: https://azure.microsoft.com/services/private-link/	You need to provide private access to services hosted on the Azure platform.
Azure Front Door: https://azure.microsoft.com/services/frontdoor/	You need to provide a security-enhanced, scalable delivery point for web applications that are global and microservice-based.
Azure CDN: https://azure.microsoft.com/services/cdn/	You need to accelerate the delivery of high-bandwidth content to global customers.
Azure Internet Analyzer (preview): https://azure.microsoft.com/services/internet-analyzer/	You need to test how performance will be impacted by networking infrastructure changes.

Table 4.2: Recommended Azure services for common use cases

Another key component of the Azure infrastructure is storage, which we will discuss in the next section.

Storage

Data is growing at a rapid rate in today's businesses. Organizations demand the best storage solutions available. When it comes to storage solutions, Azure offers a myriad of services that organizations can choose from to cater to any business cost and performance requirements:

- Azure Disk Storage
- Azure Files
- Azure Blob Storage
- Azure Data Lake Storage
- Azure NetApp Files

Let's take a closer look at each of these Azure storage offerings.

Azure Disk Storage

Azure Disk Storage is designed to be used with Azure VMs. It offers high performance as well as highly durable block storage for your mission- and business-critical applications.

Some of the key features and benefits of Azure Disk Storage include:

- **Cost-effective storage**:
  - Optimizes costs and gets the precise storage you need for your workloads with a range of disk options at a variety of price points and performance characteristics.
  - Migrate Windows and Linux-based clustered or high-availability applications cost-effectively to the cloud with shared disks.
- **Unmatched resiliency**:
  - Azure offers an availability guarantee for all single-instance VMs using Azure Disk Storage.
  - Azure Disk Storage delivers enterprise-grade durability with a zero-percent annualized failure rate.
- **Seamless scalability and highly performant**:
  - Dynamically scale performance with Azure Ultra Disk Storage and achieve high IOPS and throughput along with consistent sub-millisecond latency.
  - Scale performance on demand with performance tiers and built-in bursting capabilities to meet business demands.
- **Built-in security**:
  - Protect your data with automatic encryption using Microsoft-managed keys or your own custom keys.
  - Restrict the export and import of disks to only occur within your private virtual network with Azure Private Link support.

There are four types of disk storage options that you can choose from based on your cost and performance requirements:

- Ultra Disk Storage
- Premium SSD
- Standard SSD
- Standard HDD

Workloads that are data-intensive and transaction-heavy can benefit from the high performance and consistent low latency capabilities of **Azure Ultra Disk Storage**. Some examples of these types of workloads include SAP HANA, top-tier databases such as SQL Server and Oracle, and NoSQL databases such as MongoDB and Cassandra.

To learn more about each of these disk storage options, please visit https://docs. microsoft.com/azure/virtual-machines/disks-types.

Azure Files

Azure Files is an ideal fully managed cloud file share service that is both serverless and secure. You can access shared files on Azure Files using industry-standard NFS and SMB protocols. Furthermore, it can serve as persistent shared storage for containers with its tight integration with AKS. If you are looking for a cross-platform hybrid experience, you will be happy to know that Azure file shares can be mounted concurrently by Azure or on-premises whether you are on Windows, Linux, or macOS via Azure File Sync.

To learn more about Azure Files, please visit https://docs.microsoft.com/azure/ storage/files/storage-files-introduction.

Azure Blob Storage

Azure Blob Storage provides highly scalable, available, and secure storage. In terms of security, data can be protected using encryption at rest and advanced threat protection, while data access can be protected using **Azure Active Directory (Azure AD)**, **role-based access control** (RBAC), and network-level control.

It is ideal for the following types of workloads:

- Cloud-native and mobile applications
- High-performance computing
- Machine learning workloads
- Storage archives

Some of the key features and benefits of Azure Blob Storage include:

- **Fully supports cloud-native application development:**
  - Provides scalability and security required by cloud-native applications.
  - Supports Azure Functions and many popular development frameworks such as .NET, Python, Java, and Node.js.
- **Store petabytes of data:**
  - Archive seldomly accessed data in the most cost-effective way.
  - Replace old magnetic tapes with Azure Blob Storage and alleviate the need to migrate across hardware generations.
- **Scale up for high-performance computing (HPC):**
  - Capable of meeting the high-throughput requirements of HPC applications.
- **Scale out for billions of IoT devices:**
  - Provides storage to collect data points from billions of IoT devices.

You can learn more about Azure Blob Storage at https://docs.microsoft.com/azure/storage/blobs/storage-blobs-overview.

Azure Data Lake Storage

Azure Data Lake Storage offers a highly scalable and cost-effective data lake solution for big data analytics. It is optimized for analytics workloads and supports Azure Synapse Analytics and Power BI by providing a single storage platform for ingestion, processing, and visualization.

It is cost-effective as it allows the scaling of storage and compute to be done independently (this is something that cannot be achieved with on-premises data lakes). Azure Data Lake Storage Gen2 also allows you to further optimize costs using automated lifecycle management policies and can automatically scale up or down based on usage.

In terms of security, it provides the same level of security offered by Azure Blob Storage, including encryption at rest and advanced threat protection. Like Azure Blob Storage, data access can be protected using **Azure AD**, **RBAC**, and network-level control.

You can learn more about Azure Data Lake Storage at https://docs.microsoft.com/azure/storage/blobs/data-lake-storage-introduction.

Azure NetApp Files

Running performance-intensive and latency-sensitive file workloads in the cloud can be challenging. **Azure NetApp Files** makes it easy for enterprise **line-of-business (LOB)** and storage professionals to migrate and run complex, file-based applications with no code change. Azure NetApp Files is a fully managed Azure service powered by NetApp's industry-leading storage technology. It is widely used as the underlying shared file-storage service in various scenarios, including:

- Migration (lift and shift) of POSIX-compliant Linux and Windows applications
- SAP HANA
- Databases
- HPC infrastructure and apps
- Enterprise web applications

You can learn more about Azure NetApp Files at https://docs.microsoft.com/azure/azure-netapp-files/.

This wraps up our quick tour of Azure infrastructure with an overview of the core services of Azure: compute, networking, and storage. In the next section, we will look at some common workload migration scenarios.

Workload migration scenarios

In the previous section, we built our foundational understanding of the underlying infrastructure of Azure. We will continue to build on this knowledge by discussing the following common workload migration scenarios:

- Windows Server and Linux workloads
- SQL Server
- Containers
- VMware workloads
- SAP workloads
- HPC

Let's begin with Windows Server and Linux workloads.

Windows Server and Linux workloads

When it comes to migrating Windows Server and Linux VMs from on-premises to Azure, Microsoft recommends that you use the **Azure Migrate** hub of tools and services. Azure Migrate provides a central hub for discovery, assessment, and migration to Azure. The **Server Migration** tool in Azure Migrate is specifically built for server migration to Azure.

Some of the key features and benefits of Azure Migrate include:

- You can migrate across different environments and scenarios including on-premises servers, VMs, databases, web apps, and virtual desktops.
- You can conduct comprehensive discovery, assessment, and migration capabilities using built-in Azure tools as well as third-party tools of your choice.
- You have full visibility of your migration progress from the dashboard.
- You can migrate efficiently with rapid lift and shift migration tools at no additional cost (included as part of your Azure subscription).

Once you migrate your VMs to Azure, you should enable **Azure Automanage** to simplify mundane and repetitive IT management tasks. Azure Automanage allows IT administrators to manage and automate their day-to-day operations and perform lifecycle management across Windows and Linux servers using a point-and-click interface.

Azure Automanage works with any new or existing Windows Server or Linux VM on Azure. It automatically implements VM management best practices as defined in the Microsoft Cloud Adoption Framework. Azure Automanage eliminates the need for service discovery, enrollment, and configuration of VMs. For example, Azure Automanage enables customers to implement security best practices by offering an easy way to apply an operating system baseline to VMs per Microsoft's baseline configuration. Services such as Azure Security Center are automatically onboarded per the configuration profile chosen by the customer. If the VM's configuration drifts from the applied best practices, Azure Automanage detects and automatically brings the VM back to the desired configuration.

To learn more about Azure Migrate, visit https://docs.microsoft.com/azure/migrate/migrate-services-overview.

To learn more about Azure Automanage, visit https://azure.microsoft.com/services/azure-automanage.

To learn more about Azure Security Center, visit https://azure.microsoft.com/services/security-center/.

For information on the Microsoft Cloud Adoption Framework, visit https://docs.microsoft.com/azure/cloud-adoption-framework/.

SQL Server

Azure SQL is a family of secure, managed, and intelligent products that are built upon the familiar SQL Server database engine in the Azure cloud. Microsoft's data platform uses SQL Server technology and makes it available across a huge variety of different environments. This includes the public cloud as well as private cloud environments (which can be third-party hosted), and physical on-premises machines. As a result, Azure SQL destinations have a consistent experience with SQL Server on-premises, allowing you to bring your favorite skills, tools, languages, and frameworks to Azure with your applications. You can migrate your applications with ease using Azure Migrate and continue to use the resources that you are familiar with.

The following table provides a clear picture of each Azure SQL offering and its purpose so that you can make an informed decision that suits your requirements:

Azure SQL offering	Common use case
Azure SQL Database	Supporting a modern cloud application on an intelligent, fully managed database service that includes serverless compute.
Azure SQL Managed Instance	Migrating an existing on-premises application to Azure that requires 100% feature parity with a full-fledged SQL Server database engine. You do not want to look after the upgrade or maintenance of the SQL Server once your application is migrated to Azure.
SQL Server on Azure VMs	Easily lifting and shifting existing SQL Server workloads while maintaining 100% SQL Server compatibility and OS-level access. You want to have complete control over the database server as well as the underlying OS that it runs on.

Table 4.3: Azure SQL offerings and common use cases

We will go into each of these Azure SQL offerings in more detail.

Azure SQL Database

Azure SQL Database is an Azure **Platform as a Service (PaaS)** offering and is a fully managed SQL Server database engine hosted in Azure. Azure SQL Database is based on the latest stable SQL Server Enterprise Edition. It is an ideal choice for organizations that require cloud applications with current, stable SQL Server features while reducing development and marketing time requirements.

With Azure SQL Database, you can quickly scale up and down as required with no interruptions. It also provides additional features that are not available in SQL Server, including built-in high availability, intelligence, and management.

Azure SQL Managed Instance

Azure SQL Managed Instance is an Azure PaaS offering and is similar to an instance of the full-fledged Microsoft SQL Server database engine. It is ideal for migrating on-premises applications to Azure with minimal to no database changes. Azure SQL Managed Instance includes all the PaaS benefits of Azure SQL Database but adds capabilities that were previously only available in SQL Server on Azure VMs (as we will see in the following section), including a native virtual network and near 100% compatibility with on-premises SQL Server.

SQL Server on Azure VM

SQL Server on Azure VM is an Azure **Infrastructure-as-a-Service (IaaS)** offering and allows you to run the full fledged SQL Server inside a VM in Azure. SQL Server on Azure VMs is ideal for migrating an existing on-premises database using the lift and shift approach to Azure, with minimal to no database changes. With SQL Server on Azure VMs, you have complete administrative control over the SQL Server instance and the underlying OS that is running on the Azure VM. However, with this full control, you are also responsible for upgrading, maintaining, and backing up your software on the VM. This would not be an issue if your organization already has IT resources available to look after VMs.

These are some other benefits of SQL Server on Azure VM:

- You can install and host your SQL Server on Azure in either a Windows Server VM or a Linux VM.
- All the latest versions and editions of SQL Server are available for installation in an Azure VM.
- You can use an existing license for your SQL Server on Azure VM or provision a pre-built SQL Server VM image that already includes an SQL Server license. This is also true for the PaaS offerings via Azure Hybrid Benefit.
- SQL Server on Azure VMs allows you to build a system with high customizability to meet the performance and availability needs of your application
- You can use SQL Server on Azure VM to quickly spin up an environment to develop and test traditional SQL Server applications.

To wrap up this section, we will go over the key differences between these three Azure database offerings.

Key differences between Azure SQL Database, Azure SQL Managed Instance, and SQL Server on Azure VMs

The most significant difference between these three Azure database offerings is that SQL Server on Azure VMs allows you to have full control over the database engine. Both SQL Database and SQL Managed Instances are PaaS-based. This means there is no need for you to manage your own upgrades or backups for SQL Database and SQL Managed Instances. On the other hand, SQL Server on Azure VMs is IaaS-based, so you would be required to take care of your own upgrades for the operating system, database software, and backups. You should choose the appropriate Azure database offering according to your migration requirements.

Another technology that has been widely adopted by many organizations is containers.

Containers

Containers have gained tremendous popularity because they allow you to easily move your applications from one environment to another with no changes to your application. When it comes to containers, Azure provides several options, including:

- Azure Kubernetes Service (AKS)
- Azure Container Instance (ACI)
- Web App for Containers

ACI allows you to create an isolated container for small applications that do not require full container orchestration. These instances have a quick startup time and can leverage Azure virtual network and public IPs. ACI is ideal for **proof-of-concept (PoC)** applications.

To learn more about ACI, visit https://azure.microsoft.com/services/container-instances/.

AKS simplifies the deployment process for a managed Kubernetes cluster in Azure. Since much of the management responsibility is offloaded to Azure, both the complexity and operational overhead of managing Kubernetes are reduced.

Azure manages the Kubernetes masters and handles critical tasks such as maintenance and health monitoring for you. This means that you only have to manage the agent nodes. Therefore, you also only pay for the agent nodes, and not the masters, in your clusters; as a managed Kubernetes service, AKS is free. If you wish to create an AKS cluster, you can do so in the Azure portal. Alternatively, you can use the Azure CLI, or various template-driven deployment options such as **Azure Resource Manager (ARM)** templates and Terraform.

All nodes are deployed and configured for you when an AKS cluster is deployed—masters and agents. During deployment, you can also configure additional features such as **Azure AD** integration, advanced networking, and monitoring. AKS also supports Windows Server containers.

To learn more about AKS, visit https://docs.microsoft.com/azure/aks/intro-kubernetes.

With **Web App for Containers**, you can deploy and run containerized applications on Windows and Linux with ease. It offers built-in autoscaling and load balancing. You can automate your **continuous integration/continuous deployment (CI/CD)** process with GitHub, Azure Container Registry, and Docker Hub. You can also set up autoscaling to meet your workload demand. For example, you can set up scaling rules to reduce costs during off-peak hours.

To learn more about Web App for Containers, visit https://azure.microsoft.com/services/app-service/containers/.

VMware workloads

With Azure VMware Solution, you can seamlessly move VMware workloads from your on-premises environments to Azure. This allows you to continue to manage your existing VMware environments with the same tools that you are familiar with while running your VMware workloads natively on Azure.

You can learn more about Azure VMware Solution at https://azure.microsoft.com/services/azure-vmware/.

Another approach for migrating VMware workloads to Azure is by using the **Azure Migrate: Server Migration** tool. There are two ways in which you can migrate your VMware VMs to Azure:

1. Agentless migration

2. Agent-based migration

We will highlight the steps involved in each of these types of migrations below and provide links to resources where you can obtain detailed migration instructions.

Agentless migration

Here is an overview of the steps involved in an agentless migration:

1. Set up the **Azure Migrate** appliance.

2. Replicate the VMs.

3. Track and monitor the migration status.

4. Run a test migration.

5. Migrate the VMs.

6. Complete the migration.

For more information on agentless migration, visit: https://docs.microsoft.com/azure/migrate/tutorial-migrate-vmware

Agent-based migration

Here is an overview of the steps involved in an agent-based migration:

1. Prepare Azure to work with Azure Migrate:

    a. Create an **Azure Migrate** project.

    b. Verify the Azure account permissions.

    c. Set up a network that Azure VMs will join after migration.

2. Prepare for agent-based migration:

    a. Set up a VMware account. This is to enable **Azure Migrate** to discover machines for migration, and so the Mobility service agent can be installed on machines you want to migrate.

    b. Prepare a machine that will act as the replication appliance.

    c. Add the **Azure Migrate: Server Migration** tool.

3. Set up the replication appliance.

4. Replicate the VMs.

5. To check that all is working as expected, run a test migration.

6. Run a full migration to Azure.

For more information on agent-based migration, visit: https://docs.microsoft.com/azure/migrate/tutorial-migrate-vmware-agent

Companies with SAP workloads might be interested in migrating to Azure. We will discuss that in the next section.

SAP workloads

Azure is optimized for SAP workloads, and you can migrate most existing SAP NetWeaver and S/4HANA systems to Azure without any issues. Azure is extremely scalable and can provision VMs with more than 200 CPUs and terabytes of memory. For the most demanding workloads, Azure also offers **HANA Large Instances** (**HLIs**). HLIs are based on dedicated physical hardware in an Azure datacenter running Intel Optane, which is a unique cloud offering.

To deploy your SAP workloads into Azure IaaS successfully, you must understand the differences between the offerings of traditional hosting providers and Azure IaaS. A traditional hosting provider would adapt infrastructure (for example, server type, storage, and network) to the workload a customer wants to host. For Azure IaaS, the onus is on the customer to identify the requirement for the workload and to choose the appropriate Azure VMs, storage, and network components for the deployment.

To plan for a successful migration, you should make note of the following:

- What types of Azure VMs would support your SAP workload requirements?
- What Azure database services can provide the support required by your SAP workloads?
- Gain an understanding of the various SAP throughput offered by supporting Azure VM types and HLI SKUs.

Additionally, it is necessary to find out how Azure IaaS resources and bandwidth limitations stand in comparison to the actual consumption by the on-premises resource counterparts. As such, you must be familiar with the various capabilities of Azure VMs and HLI supported with SAP in terms of:

- Memory and CPU resources
- Storage IOPS and throughput
- Network bandwidth and latency

To learn more about SAP migration to Azure, visit: https://azure.microsoft.com/solutions/sap/migration/

To learn how Microsoft migrated their own SAP applications to Azure and other SAP on Azure scenarios, visit: https://www.microsoft.com/itshowcase/sap-on-azure-your-trusted-path-to-innovation-in-the-cloud

Organizations with complex computational and process-intensive tasks often require HPC resources. We will take a close look at HPC offerings for Azure in the next section.

High-performance computing

As companies acquire larger volumes of data and more sophisticated methods to manipulate it, HPC becomes more prevalent. In recent times, more and more organizations are starting to understand how they can reinvent and transform their business using HPC solutions. However, many of these organizations do not have a history in HPC, and so they have no existing hardware clusters or other investments in HPC hardware to use as a starting point. This is where Azure HPC can offer some of its greatest value. Azure HPC gives you the ability to develop and host game-changing applications and experiences without needing to deploy on-premises infrastructure.

Here are some examples of different industries and what they would use HPC for:

Industries	What do they use HPC for?
Finance	Risk modeling, fraud prevention
Engineering	**Computational fluid dynamics (CFD)**, **finite element analysis (FEA)**, **electronic design automation (EDA)**, chemical engineering simulation, autonomous vehicle development
Life sciences	Genome sequencing, DNA splicing, molecular biology, pharmaceutical development
Earth sciences	**Weather modeling (WRF)**, seismic processing, reservoir modeling
Manufacturing	Digital vehicle engineering, Industry 4.0, predictive maintenance, digital twins

Table 4.4: How HPC is used across different industries

Azure enables you to run enterprise-grade HPC workloads in the cloud, without the costs and risks of investing in planning, deploying, and managing your own HPC clusters on-premises. An Azure HPC system has the advantage that you can dynamically provision resources as they are needed and shut them down when demand falls. Azure makes it easy to coordinate an HPC task across many VMs, and it supports a wide variety of VM sizes, processer types, and storage options common to HPC data requirements.

In Azure, you have a choice of several technologies to meet your HPC demands:

- **H-Series Virtual Machines**: CPU-based VMs with high-performance interconnect
- **N-Series Virtual Machines**: GPU- and other accelerator-based VMs with high-performance interconnect
- **Cray Supercomputer**: Options for dedicated, fully managed environments
- **Azure Cycle Cloud**: For scaling up environments and clusters
- **Azure HPC Cache**: For on-/off-premises big data synchronization
- **Azure Batch**: For scaling application jobs

Some of the leading industry ISVs running on Azure include **Ansys**, **Altair**, and **Willis Towers Watson**. Let's take a closer look at each of these Azure HPC offerings.

> **Note**
>
> Linux is the most common OS for HPC workloads and thus Azure fully supports Linux as the default OS for HPC VMs.

Azure H-Series Virtual Machines

The **H-Series Virtual Machines** are optimized for HPC applications that have exceptionally high memory bandwidth and scalability demands, for example:

- Computational chemistry
- Electronic design automation
- Finite element analysis
- Fluid dynamics
- Heat transfer simulation
- Quantum simulation
- Rendering
- Reservoir simulation
- Risk analysis
- Seismic processing
- Spark
- Weather modeling

The H-Series also supports extremely fast interconnects with RDMA InfiniBand, making them highly capable for tightly coupled workloads that require a lot of communications between servers during processing.

You can learn more about Azure VM HPC instances at https://azure.microsoft.com/pricing/details/virtual-machines/series/.

Azure N-Series Virtual Machines

The **N-Series Virtual Machines** support a variety of GPUs and are suitable for compute and graphic-intensive workloads, including:

- Deep learning
- High-end remote visualization
- Predictive analytics

These VMs also support extremely fast interconnects with RDMA InfiniBand, making them highly capable for tightly-coupled workloads that require a lot of communications between servers during processing.

You can learn more about the N-Series VMs at https://azure.microsoft.com/pricing/details/virtual-machines/series/.

Cray supercomputer

Cray in Azure provides you with a dedicated, fully managed supercomputer on your virtual network. Microsoft and Cray have teamed up to offer you extreme performance, scalability, and elasticity capable of handling the most demanding HPC workloads. You can now get your own Cray supercomputer delivered as a managed service and run it alongside the other Azure services to power your big compute workflows.

You can learn more about Cray in Azure at https://azure.microsoft.com/solutions/high-performance-computing/cray/.

Azure CycleCloud

Azure CycleCloud is an enterprise tool for managing and orchestrating HPC environments on Azure. It is targeted at HPC administrators who wish to deploy an HPC environment using a particular scheduler. Out of the box, Azure CycleCloud supports many widely used HPC schedulers, including:

- Grid Engine
- HTCondor
- PBS Professional
- Platform LSF
- Slurm Workload Manager

With Azure CycleCloud, HTC administrators can:

- Automatically scale the infrastructure to run jobs efficiently at any scale.
- Create and mount different types of file systems to the compute cluster nodes to support HPC workloads.
- Provision infrastructure for HPC systems.

It is interesting to note that Azure CycleCloud and **Azure Batch** are sister products. We will talk about Azure Batch later in this section.

You can learn more about Azure CycleCloud at https://docs.microsoft.com/azure/cyclecloud/overview.

Azure HPC Cache

Azure HPC Cache brings the scalability of cloud computing to your existing workflow by caching files in Azure and by helping to improve the speed of access to your data for HPC tasks. You can even use Azure HPC Cache to access your data across WAN links, such as in your local datacenter **network-attached storage** (**NAS**) environment. We will talk more about Azure HPC Cache later in this chapter.

You can learn more about Azure HPC Cache at https://azure.microsoft.com/services/hpc-cache/.

Azure Batch

Azure Batch is a service for working with large-scale parallel and computationally intensive tasks on Azure. Unlike HPC VMs and Microsoft HPC Pack, Azure Batch is a managed service. You provide data and applications, and you specify whether to run on Linux or Windows, how many machines to use, and what rules apply to autoscaling. Azure Batch handles provisioning of the compute capacity and optimizes the way the work is done in parallel. You only pay for the underlying compute, networking, and storage you use. The Azure Batch scheduling and management service is free.

Azure Batch is an ideal service for heavy workloads, such as financial risk modeling, 3D rendering, media transcoding, and genetic sequence analysis. Think of Azure Batch as a flexible management and scheduling service layer on top of the Azure platform. Even though you might be able to provision thousands of VMs to support heavy workloads without the help of Azure Batch, doing so without Azure Batch would require you to look after all the scheduling of thousands of VMs, and for distributing the work according to available capacity yourself.

You can learn more about Azure Batch at https://docs.microsoft.com/azure/batch/batch-technical-overview.

The HPC services in Azure put HPC techniques at your fingertips and empower you to perform new tasks. You've learned about the solutions available on Azure for HPC workloads: Azure Batch, HPC VMs, and the Microsoft HPC Pack. You can now choose the best option for your HPC workloads.

This concludes our overview of common workload migration scenarios. In the next section, we will look at how to achieve cloud scale and maximize performance in Azure.

Achieving cloud scale and maximizing performance

Here, we will look at how to achieve cloud scale and maximize performance once you have migrated your workload to Azure. We will start by looking at Azure scale and how autoscaling can help achieve this goal. We will then discuss the options available for compute and storage performance optimization.

Azure autoscaling

Autoscaling enables a system to adjust the resources required to meet the varying demand from users while controlling the costs associated with these resources. You can use autoscaling with many Azure services, such as Azure **Virtual Machine Scale Sets**. Autoscaling requires you to configure autoscale rules that specify the conditions under which resources should be added or removed.

Suppose you run an online specialty store. During holiday seasons, your website might experience a short-term increase in traffic. Such spikes can happen at any time, which makes it difficult to plan for potential spikes in traffic. This unpredictability of events means manual scaling is not an option and it would be prohibitively expensive to keep your website resources available at all times in case of a sudden spike. This is where Azure autoscaling comes in handy. It will automatically scale up or down your resources according to your configured autoscale rules, giving you peace of mind.

In a similar vein to scaling, we also need to look at compute and storage performance considerations.

Compute and storage performance considerations

As you may recall, in the *Choosing your underlying infrastructure* section of this chapter, we highlighted various Azure services for compute and storage. When it comes to compute performance considerations, Azure **Virtual Machine Scale Sets** is a service that offers both high performance and high availability and can be used instead of availability sets. For your storage performance consideration, Azure Ultra Disk provides the highest performance with the lowest latency. In this section, we will dive deeper into each of these two services for your compute and storage performance considerations.

Compute performance consideration: Virtual Machine Scale Sets

With Virtual Machine Scale Sets, you can create and manage a diverse group of load-balanced VMs and automatically scale up and down the number of VMs based on actual demand and usage, or as per a customized schedule defined by you. Best of all, you do not need to manually configure each VM individually.

Under the hood, Virtual Machine Scale Sets use a load balancer to distribute requests across the VM instances, and a health probe to determine the availability of each instance. This is how it works:

1. Virtual Machine Scale Sets use the health probe to ping the instance.

2. If the instance responds, the Virtual Machine Scale Set determines that the instance is still available.

3. If the ping fails or times out, the Virtual Machine Scale Set determines that the instance is unavailable and stops sending requests to it.

Virtual Machine Scale Sets support both Linux and Windows VMs in Azure and allow you to centrally manage, configure, and update a heterogeneous group of VMs. As demand grows, the number of VMs running in the scale set increases. Conversely, as demand drops, excess VMs can be shut down. You can have up to 1,000 VMs on a single Virtual Machine Scale Set per Availability Zone.

If you deal with large workloads whose demand varies and is unpredictable, Virtual Machine Scale Sets are an ideal solution. You can get full control of individual VMs within your scale set and ensure high availability at scale with the flexible orchestration mode. You can change VM sizes without redeploying your scale set or mix **Azure Spot VMs** and pay-as-you-go VMs within the same scale set to optimize your costs. You can also manage VMs and Virtual Machine Scale Sets using the same APIs and accelerate your deployments by defining fault domains during the VM creation process.

To learn more about Azure Virtual Machine Scale Sets, visit https://azure.microsoft. com/services/virtual-machine-scale-sets/.

Having discussed compute, now we will look at what storage options you should be considering for HPC workloads.

High-performance file caching consideration: Azure HPC Cache

As discussed earlier, **Azure HPC Cache** speeds up access to your data for HPC tasks by caching files in Azure. From the Azure portal, you can easily launch and monitor Azure HPC Cache. Even if you alter the storage target at the back end, client access remains simple, since new blob containers or existing NFS storage can become parts of Azure HPC Cache's aggregated namespace.

To find out more, you can watch a short video about Azure HPC Cache at https://azure. microsoft.com/resources/videos/hpc-cache-overview/.

Latency-sensitive storage consideration: Azure NetApp Files

Azure NetApp Files is an enterprise-class, metered file storage service that lets you choose the service and performance levels you want. By default, it is highly available, high-performance, and all workload types are supported. The service allows you to set up snapshots on demand, as well as to manage policies (currently in preview) that schedule automatic volume snapshots.

Azure NetApp Files makes it easy for enterprise LOB and storage professionals to migrate and run complex, file-based applications with no code change. It is the shared file-storage service of choice that underlies a wide range of scenarios, including the migration of POSIX-compliant Linux and Windows apps (lift and shift), enterprise web apps, databases, SAP HANA, and HPC apps and infrastructure.

Storage performance consideration: Azure Ultra Disks

As discussed earlier in this chapter, if your Azure VM requires the highest performance, throughput, and IOPS with the lowest latency, you should consider Azure Ultra Disk. Azure Ultra Disk provides top-tier performance at the same availability levels as existing disk offerings. One major benefit of Azure Ultra Disk is that you can dynamically adjust the SSD performance along with your workload without having to restart your Azure VMs. This makes it ideal for heavy workloads that are transaction-intensive such as SAP HANA, SQL Server, and Oracle.

To learn more about Azure Ultra Disks, visit https://docs.microsoft.com/azure/virtual-machines/disks-enable-ultra-ssd.

In the next section, we will talk about business continuity and disaster recovery, and how to accomplish them using the appropriate Azure tools.

Enterprise-grade backup and disaster recovery

At the time of writing, and mostly due to the pandemic, organizations of all sizes have opted to migrate their operations to Microsoft Azure, to enable secure remote work and improve operational efficiency. In addition, organizations that migrate to Azure can significantly reduce CapEx, since there is very little or no infrastructure at all that is acquired in an initial investment when deploying workloads in the public cloud. However, it is important to design a reliable infrastructure that can help you maintain your data and applications so they are always available and accessible.

In order to help you protect your data and critical applications from any potential failure, it is important to always make copies of your data and provide secure access to those backups to ensure business continuity. Azure Backup can help you back up your data and restore it in Azure.

Azure Backup

Performing backups in the cloud can help organizations reduce costs and improve consistency in storage management. Azure Backup is designed to help you back up and restore your data to the cloud.

While it is known that Azure takes at least three copies of your data and stores it using Azure storage, it is critical to keep a backup and improve your security posture by enabling protection against corruptions, accidental deletions, or ransomware.

Since Azure Backup is a fully managed service, built into Azure, you can make backups of data that is sitting on-premises or in the cloud and restore it back to a specific period to ensure business continuity.

Azure Backup doesn't require you to set up any infrastructure, which makes it simple to use while reducing the cost of ownership. Azure Backup provides native integration with different workloads that run on Azure, such as Azure VMs, databases like SQL Server on virtual machines, or Azure PostgreSQL, SAP, and Azure files, so there's no need to provision any infrastructure to perform backups.

So, how does Azure Backup store the data? Under the hood, Azure Backup relies primarily on Azure Blob Storage to store your backups and ensure reliability as you can choose from a variety of redundancy options for your backups, like **locally redundant storage (LRS)**, **geo-redundant storage (GRS)**, **read-access geo-redundant storage (RA-GRS)**, or **zone-redundant storage (ZRS)**. Azure Backup improves security, as it has encryption capabilities to secure data in transit and at rest, and features like role-based access control as well as providing soft delete for backup data for up to 14 days with no additional charges.

As organizations' data increases along with their workloads across multiple Azure subscriptions, Azure regions, and even tenants, it is critical to keep their data and resources secure and compliant. Therefore, governance features become relevant to monitor and enforce governance standards on the backups.

To learn more about guidance and best practices for Azure Backup, visit https://docs.microsoft.com/azure/backup/guidance-best-practices.

Backup Center is a new and native central management capability that helps monitor, operate, govern, and get insights into your backup across all your backup estate. You can manage all the data sources and backup instances across all your vaults. You can also select a specific data source to get more details about the backup. Through Backup Center, you can initiate restores of your data and add policies and vaults as shown in *Figure 4.1*:

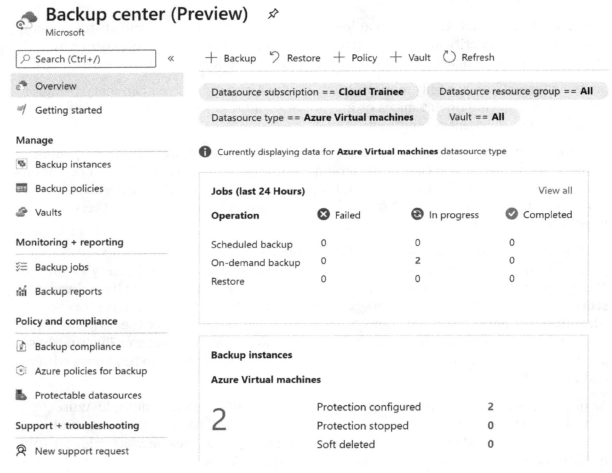

Figure 4.1: Backup Center page

Using Backup Center, you can integrate governance features such as Azure policies that allow organizations to audit and deploy policies to reach a desired backup goal state. Then, you can use backup compliance to see if your organization is adhering to those policies.

Now that we have reviewed how Azure Backup works, another consideration for building reliable systems is the recovery capacity. Let's see how Azure can help you improve failover and recovery processes through Azure Site Recovery.

Azure Site Recovery

As an organization, you need to ensure business continuity and reduce downtime when a disaster occurs. Planning for business continuity and disaster recovery implies the adoption of a mechanism to ensure your workloads and data are safe and resilient to planned or unplanned outages.

Azure Site Recovery (ASR) can help you keep your applications and workloads running during planned or unplanned outages. ASR has the capacity to replicate workloads that are running on-premises or on VMs and restore them in a different location. This way, if an outage occurs at your primary location, you will have the ability to replicate your workloads to a secondary location and be able to ensure business continuity. Once your primary location is up and running, you can then perform a failback to it.

With such a mechanism for disaster recovery, ASR helps in reducing infrastructure costs by eliminating the need for building or maintaining a costly secondary datacenter. Plus, you only pay for compute resources when VMs are spun up, meaning at the time of actual failover.

Furthermore, it is simple to deploy and manage. There is a point-and-click graphical UI for setting up replication and conducting ongoing operations, including no-impact DR drills. This means the entire disaster recovery plan can be tested by failing over to a secondary site, without impacting the production site. And finally, there is seamless integration with other Azure services.

Let's take an example of an organization that is looking to enable a disaster recovery scenario for their applications running on VMs on Azure. When enabling ASR for a VM, ASR requires the installation of an extension on the VM—this extension is the Site Recovery Mobility service as shown in *Figure 4.2*:

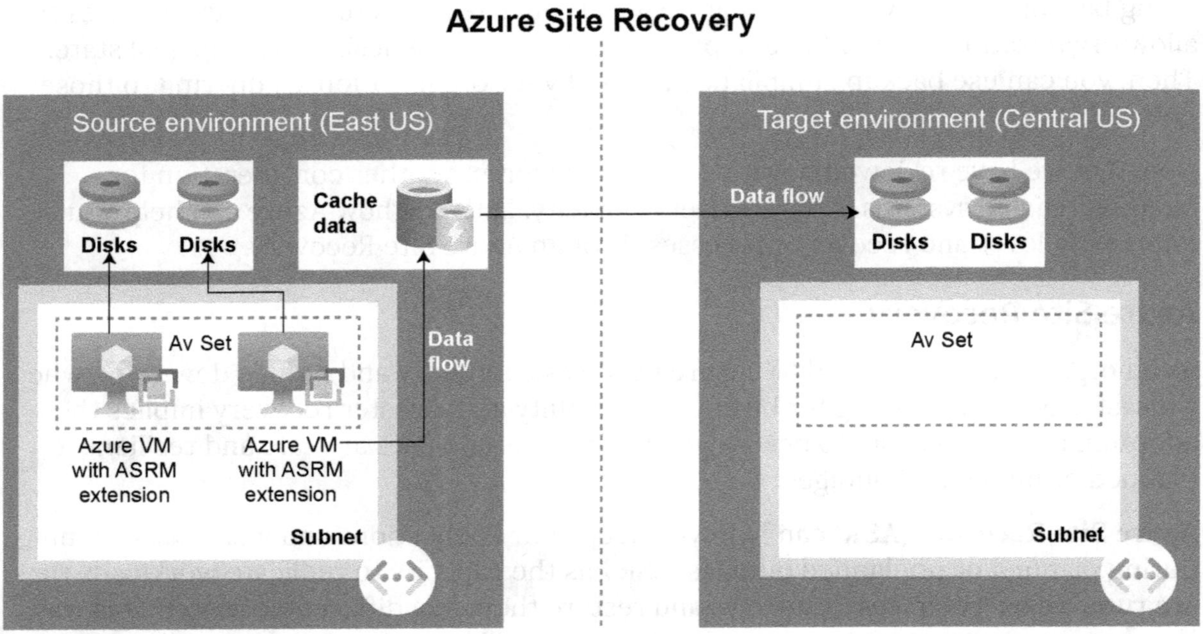

Figure 4.2: Azure Site Recovery

> **Note**
>
> The Site Recovery Mobility service must be installed on the VM that is to be replicated. If you plan to enable ASR on an Azure **Network Virtual Appliance (NVA)**, we strongly recommend you validate whether the extension can be installed, as many NVAs have a closed operating system, and it is not possible to enable ASR for those virtual machines.

During the replication process, VM disk writes are sent to a cache storage account in the source region, and data is sent from the cache storage account to the target region or secondary region. At this moment, Azure establishes recovery points from the data replicated.

When performing a failover for the VM, the recovery point is utilized by ASR to restore the VM to the target region.

Overall, to enable ASR on your VMs, you must:

- **Create a Recovery Services vault** to store the data that is going to be copied along with the configuration information for the VMs. It is recommended to have at least a contributor role in the subscription.
- **Enable replication** and **configure the source and target settings** when performing a failover. All resources, including the VMs and networking components, are created in the target region.
- **Prepare the VMs** to ensure outbound connectivity and verify that the VMs have the root certificates installed.
- **Run a test failover**, and we suggest using a non-production network for testing purposes to avoid any impact on the resources in the production network.

Note that access to storage accounts that are utilized by ASR vaults is only allowed through Azure Backup. This way you can improve security and protect your data.

This wraps up our discussion on business continuity and disaster recovery. In the final section, we will go through some useful resources to guide you through your cloud migration journey in Azure.

Azure migration best practices and support

You can browse the following resources for documentation, tutorials, and more, depending on your needs.

The official Azure documentation

The official Azure documentation provides a wealth of information on how to get started, how to architect and design your applications for the cloud, documentation on all Azure products, tutorials, and more: https://docs.microsoft.com/azure/.

Azure Migrate

Learn how you can use Azure Migrate to discover, assess, and migrate your on-premises infrastructure, data, and applications to Azure: https://docs.microsoft.com/azure/migrate/migrate-services-overview.

Azure Migration Program

Learn how the Azure Migration Program can accelerate your cloud migration journey with best practices, resources, and guidance: https://azure.microsoft.com//migration/migration-program/.

Azure Architecture Center

Learn how to design applications on Azure that are secure, resilient, scalable, and highly available by following proven practices from the industry: https://docs. microsoft.com/azure/architecture/guide/.

Azure App Service

Learn how to use Azure App Service with quickstarts, tutorials, and examples: https:// docs.microsoft.com/azure/app-service/.

Azure SQL Database

Find more about the Azure SQL family of SQL Server database engine products in the cloud: https://docs.microsoft.com/azure/azure-sql/.

Free training with Microsoft Learn

To ensure that teams gain and maintain proficiency in Azure, Microsoft Learn provides a free training platform for attaining new skills and certifications through interactive online learning: https://docs.microsoft.com/learn/.

Contact an Azure sales specialist

Speak to Microsoft's team of specialists to see how Microsoft can help you migrate to Azure: https://azure.microsoft.com/migration/web-applications/app-migration-contact-sales/.

Summary

Throughout this chapter, we reviewed how you can plan, design, and implement reliable systems in the cloud. We went through the Microsoft Azure Well-Architected initiative and how it can help you adopt best practices to improve the quality of your workloads running in the cloud. Then we took a deep dive into the underlying infrastructure of Azure to give you all the information you need to decide how best to meet your migration goals. We also talked about common workload migration scenarios and how to achieve cloud scale and maximize performance on Azure.

Lastly, we reviewed how ASR and Azure Backup can help you quickly pivot to remote work for your employees, ensuring business continuity during a failover and access to your data in a secure manner.

In the next chapter, we will review how you can improve your security posture when enabling remote work.

5

Enabling secure, remote work with Microsoft Azure AD and WVD

In the previous chapter, we learned that transitioning to the cloud isn't just about moving on-premises resources to the cloud, but it is a journey that involves improving scalability, security, and cloud-based productivity to enable remote collaboration work from anywhere.

In this chapter, we will address common infrastructure challenges that organizations face when trying to enable remote work, and how Azure core infrastructure components such as networking, identity, security, compute, and storage can empower today's modern digital workspace. In this chapter, we will cover:

- The fundamentals of deploying remote work infrastructure
- Azure Active Directory
- Enabling remote connectivity
- Securing and managing traffic to your workloads
- Enabling remote workers with Windows Virtual Desktop

We'll begin by talking about the fundamentals of deploying remote work with Azure.

The fundamentals of deploying remote work infrastructure

Remote working has been an imperative transition for most organizations in the past year and has brought multiple challenges for people who used to work from corporate offices full-time.

Enabling working from home is the key to helping people stay connected across an organization and with partners and customers. By leveraging **Azure Active Directory (Azure AD)**, not only will you be able to enable remote working but you will also be able to allow users to securely access the applications they need from outside the corporate network.

It is important to have a strong identity foundation in the cloud, and Azure AD is the key component to empower users working remotely. Azure AD is a cloud-based offering from Microsoft that provides identity and access management capabilities. This means that Azure AD can span across different environments, allowing employees access to resources that can be located in their actual on-premises datacenter or cloud-based applications.

Azure AD is the largest cloud-based identity service provided in the world, with over 200,000 organizations and more than 30 billion daily authentication requests. Azure AD also helps you detect and remediate security threats. One of the main challenges for organizations looking to use identity and access management services includes the onboarding process.

Let's take a quick look at the key components and configurations that can help you securely enable application access.

Azure Active Directory

Before we review Azure AD's capabilities for enabling remote work, it is important to understand that Azure AD is a cloud-based service for identity and access management. Therefore, it is suitable for different types of users, such as:

- IT admins looking to control access to applications while protecting user identities and credentials to meet governance requirements.

- App developers who can use Azure AD's capabilities and integration with Azure AD APIs to provide personalized app experiences.

- Subscribers to multiple services such as Microsoft 365, Dynamics 365, or Microsoft Azure.

There are some considerations that you should take into account when choosing Azure AD, which we will cover next.

Plan your environment

Prior to performing any deployment for Azure AD, we need to ensure that we have the right licensing model that will allow us to properly configure Azure AD with the features needed for our organization. There are three main Azure AD licensing tiers:

- Azure AD Free

- Azure AD Premium P1

- Azure AD Premium P2

Starting with **Azure AD Free**, this tier will provide you with some basic features such as user and group management, on-premises directory synchronization, self-service password change for cloud users, and single sign-on across cloud platforms like Azure, Microsoft 365, and third-party SaaS solutions, as seen in *Figure 5.1*:

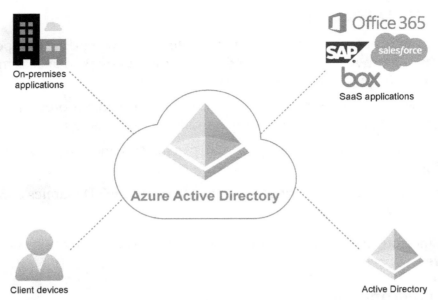

Figure 5.1: Azure AD

If you're looking to allow users that are part of your organization access across both on-premises and cloud resources, then **Azure AD Premium P1** could fit your organization's primary needs. This tier supports dynamic groups and self-service group management, and it enables Microsoft Identity Manager and cloud write-back capabilities to allow your on-premises users to perform a self-service password reset, as shown in *Figure 5.2*:

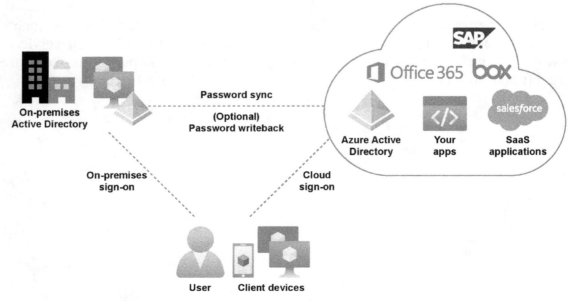

Figure 5.2: Azure AD hybrid environment

Azure AD Premium P2 is the ideal tier if your organization is looking to take advantage of advanced security capabilities such as risk-based access policies based on Azure AD identity protection, which can detect malicious activities by using cloud-scale machine learning algorithms, as shown in *Figure 5.3*:

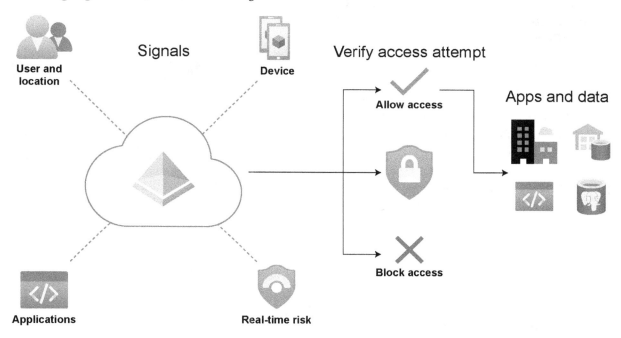

Figure 5.3: Azure AD conditional access

When your organization signs up for a Microsoft service such as Azure, Microsoft Intune, or Microsoft 365, it will be entitled to a dedicated Azure AD service instance called a tenant. A tenant represents an organization.

You can leverage the creation of your Azure AD tenant using the Azure portal, in which you can configure and do all the administrative tasks related to Azure AD. In addition, you can make use of the Microsoft Graph API to access Azure AD resources.

Now that we have gone through the various scenarios and licensing models available for Azure AD, we will proceed to review the configuration of the actual deployment of the Azure AD tenant.

Configuring Azure AD and hybrid environments

While you might be familiar with most of the **Active Directory Domain Services (AD DS)** capabilities, it is important to highlight that Azure AD and AD DS have some similarities and differences, and it is worth performing a quick review to better understand the scope of each product. The following table summarizes some of the core differences:

Feature	Active Directory Domain Services (AD DS)	Azure Active Directory (Azure AD)
Provisioning users	You can create internal users or use a provisioning system.	You can extend your organization identity infrastructure to the cloud through Azure AD Connect to sync identities.
Enabling external identities	You can create external users manually as regular users in a dedicated external AD forest.	Azure AD provides support for external identities through Azure AD B2B, including local, enterprise, or social account identities, enabling single sign-on access to your applications and APIs.
Credentials	You can manage credentials **based on passwords, certificate** authentication, and smartcard authentication.	Azure AD spans across cloud and on-**premises and significantly improves security** using multi-factor authentication and passwordless technologies, such as FIDO2.
Infrastructure	Active Directory can enable the on-premises infrastructure components such as DNS, DHCP, IPSec, WiFi, NPS, and VPN access.	Azure AD is the control plane for accessing apps, and you can have control over users that should or should not have access to apps under required conditions.
Legacy applications	On-premises apps use LDAP, NTLM and Kerberos or Header-based authentication to control access to users.	Azure AD can allow remote workers to access on-premises apps using Azure AD Application Proxy agents running on-premises.
Windows desktops	You can join Windows devices and manage them using Group Policy **and System Center Configuration** Manager.	You can use Azure AD to join Windows devices and leverage conditional access to apply the right access controls and enforce organizational policies.
Windows servers	Management capabilities for on-premises Windows servers can include Group Policy or other management solutions.	You can use Azure AD DS to manage your Windows Server machines on Azure.

Table 5.1: Comparison between AD DS and Azure AD

Azure AD managed identities can be used if your organization has Linux workloads and needs to provide secure access and communication to those resources.

In order to start working with Azure AD, you are required to use a tenant, which usually represents an organization. It is possible to start working with an existing tenant, or you can create a new Azure AD tenant.

When you create an Azure subscription, it will have a trust relationship with Azure AD to authenticate users, services, and devices. Your organization can have multiple Azure subscriptions, establish a trust relationship with an instance of Azure AD, and leverage a security principal to provide access to the resources that are secured by the Azure AD tenant.

Based on the types of users your application is going to authenticate, there are two main types of environments that can be created:

- Azure AD accounts (or Microsoft accounts)
- Azure AD B2C accounts

The Azure AD accounts, or Microsoft accounts, refer to accounts such as outlook.com and live.com or work- and school-related accounts. The Azure AD B2C accounts refer to using local or social accounts such as your social identity for Facebook or Twitter, among other identities, to get single sign-on access to your applications and APIs.

Creating your Azure AD tenant

The creation of a new Azure AD tenant can be done through the Azure portal. Once you have your Azure subscription ready to use, the next step is to go to the Azure portal, then select the **Azure Active Directory** option, as shown in *Figure 5.4*:

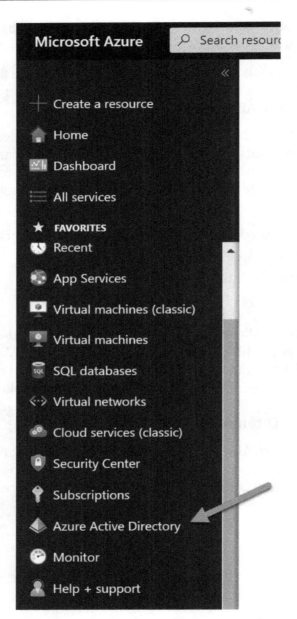

Figure 5.4: Azure AD in the Azure portal

On the overview page, click on **Create a tenant**, as shown in *Figure* 5.5:

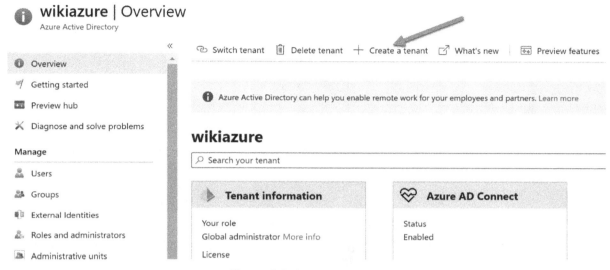

Figure 5.5: Create a tenant

Then, on the **Basics** pane, you can choose the type of tenant you want to create. Select **Azure Active Directory** or **Azure Active Directory (B2C)**, as shown in *Figure* 5.6:

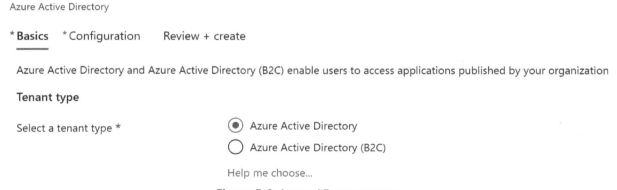

Figure 5.6: Azure AD tenant type

In the **Configuration** pane, you can provide a name for the organization and the initial domain name you want to use, as shown in *Figure* 5.7:

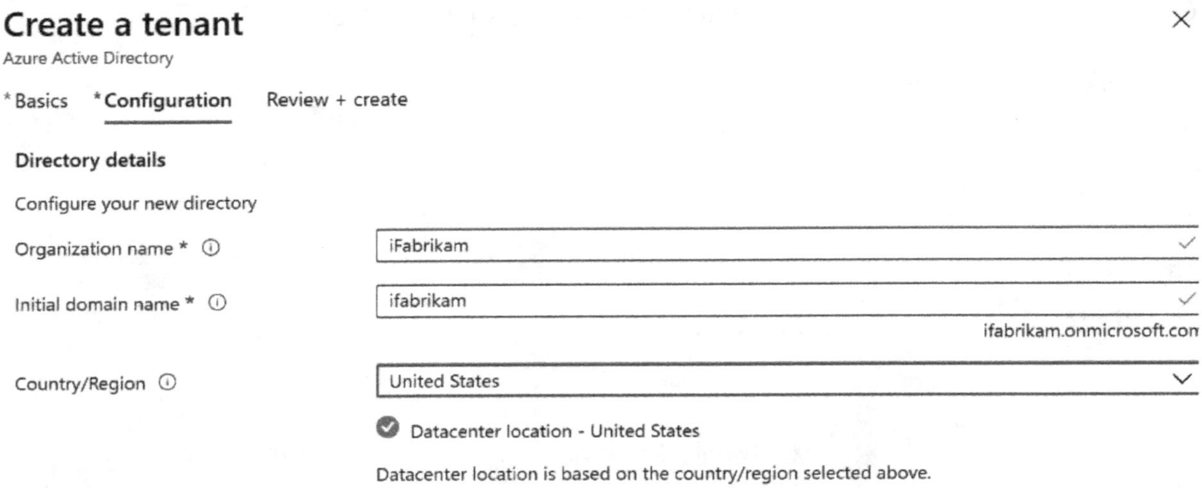

Create a tenant

Azure Active Directory

* Basics   * **Configuration**   Review + create

Directory details

Configure your new directory

Organization name * ⓘ

| iFabrikam | ✓ |

Initial domain name * ⓘ

| ifabrikam | ✓ |

ifabrikam.onmicrosoft.com

Country/Region ⓘ

| United States | ⌄ |

✅ Datacenter location - United States

Datacenter location is based on the country/region selected above.

Figure 5.7: Azure AD configuration

Next, Azure is going to validate your deployment. Once it is complete, click on **Create**, as shown in *Figure 5.8*:

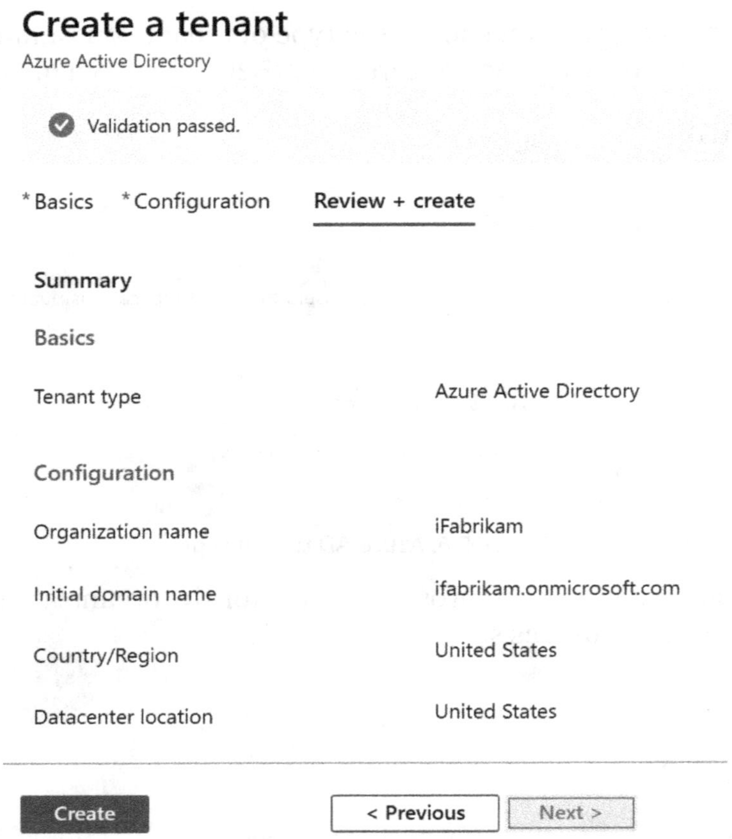

Create a tenant

Azure Active Directory

✅ Validation passed.

* Basics   * Configuration   **Review + create**

Summary

Basics

Tenant type   Azure Active Directory

Configuration

Organization name   iFabrikam

Initial domain name   ifabrikam.onmicrosoft.com

Country/Region   United States

Datacenter location   United States

[Create]   [< Previous]   [Next >]

Figure 5.8: Creating the Azure AD tenant

You should see a notification in the Azure portal related to the creation of the Azure AD tenant, as shown here:

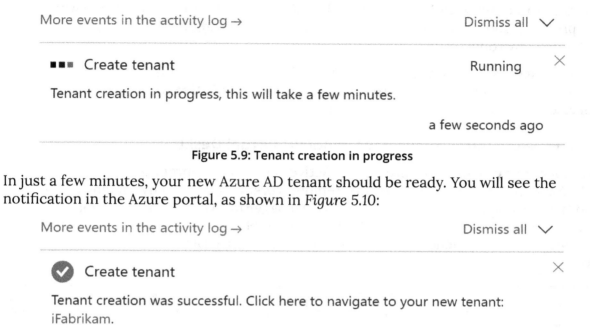

Figure 5.9: Tenant creation in progress

In just a few minutes, your new Azure AD tenant should be ready. You will see the notification in the Azure portal, as shown in *Figure 5.10*:

Figure 5.10 Azure AD tenant creation successful

Click on the notification to be redirected to your new tenant page, as shown in *Figure 5.11*:

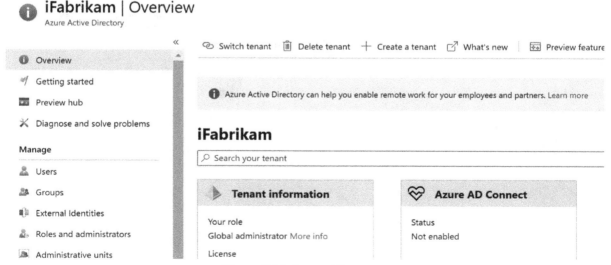

Figure 5.11: Azure AD tenant

For hybrid environments, it is recommended that you use Azure AD Connect to achieve hybrid identity integration, as it provides the features required to synchronize users on-premises with Azure AD, but also allows users to use the same password on-premises and in the cloud without the need for the additional infrastructure of a federated environment.

Although federation is an optional configuration, Azure AD Connect can be helpful for hybrid configurations that are using an on-premises **Active Directory Federation Services (AD FS)** infrastructure.

Preparing the infrastructure for Windows Virtual Desktop

Whether your organization already has Active Directory on-premises or is just starting with Azure AD, it is possible to enable remote working through a hybrid environment and take advantage of the **Azure Resource Manager (ARM)** templates to simplify the creation of your cloud environment. The ARM template shown in *Figure 5.12* will assist you with the deployment of a virtual machine with Active Directory and Azure AD Connect installed:

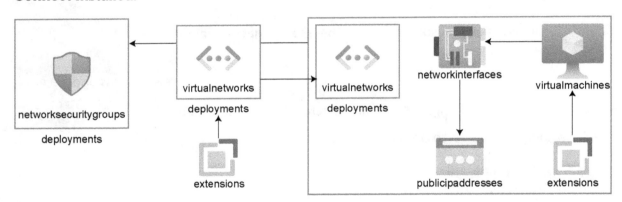

Figure 5.12: ARM template components

This ARM template will provision the following resources:

- One virtual network
- One subnet
- One **Network Security Group (NSG)**:
  - Permits AD traffic, permits **Remote Desktop Protocol (RDP)** incoming traffic, and restricts **Demilitarized Zone (DMZ)** access
- DNS configured to point to the domain controller
- One virtual machine:
  - AD DS is installed and configured.
  - Test users are created in the domain.
  - Azure AD Connect is installed and ready for configuration.
- One public IP address assigned for remote administration via RDP

Once the ARM template is deployed, the status will change to **complete**. At this point, the domain controller is ready for RDP connectivity.

> **Note**
>
> You can also try different ARM templates based on your needs and pull them from the Azure Quickstart templates, such as:
>
> active directory new domain HA 2 DC zones: https://github.com/Azure/azure-quickstart-templates/tree/master/active-directory-new-domain-ha-2-dc-zones
>
> active-directory-new-domain-module-use: https://github.com/Azure/azure-quickstart-templates/tree/master/active-directory-new-domain-module-use
>
> active-directory-new-domain: https://github.com/Azure/azure-quickstart-templates/tree/master/active-directory-new-domain

Now we can proceed to configuring Azure AD Connect with AD DS. You can connect to the Azure virtual machine recently provisioned to install Azure AD Connect using this link: https://www.microsoft.com/download/confirmation.aspx?id=47594

Configuring Azure AD Connect with AD DS

As mentioned previously, Azure AD Connect will help you achieve hybrid identity integration as it provides features to synchronize users on-premises.

The next step in the process is to connect to the domain controller and configure Azure AD Connect. If your organization has a single-forest topology, then you can use the **Express Settings**.

You can find more details on the installation of Azure AD Connect here: https://docs. microsoft.com/azure/active-directory/hybrid/how-to-connect-install-custom.

The following figure shows the Azure AD Connect wizard:

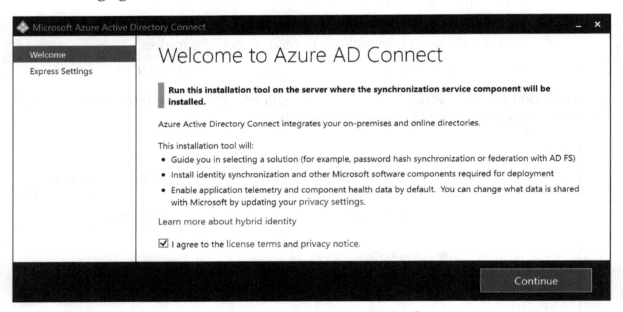

Figure 5.13: Azure AD Connect wizard

At this point, we have reviewed Azure AD Connect with AD DS in order to synchronize with Azure AD. Once your organization has Azure AD properly configured, it is important to protect employees' identities with stronger authentication mechanisms and minimize potential identity attacks.

Secure your identity infrastructure

We live in a world where passwords are no longer good enough and the majority of attacks happen to be related to password weaknesses and data breaches, most of them involving phishing. It is important to protect your identity with stronger authentication mechanisms. Azure AD **Multi-Factor Authentication (MFA)** is one of them—let's have a look.

Azure AD Multi-Factor Authentication

Strong authentication using Azure AD can prevent up to 99.9% of identity attacks. As you develop your strategy for strengthening your credentials, you should deploy the most secure, usable, and cost-effective methods for your organization and your users. One of the main benefits of using Azure AD is the support for MFA, which requires the use of at least two verification methods. This way, Azure AD MFA helps secure access to data and applications.

Azure AD MFA capabilities are available and licensed in different ways. You can refer to the documentation for more details: https://docs.microsoft.com/azure/active-directory/authentication/howto-mfa-mfasettings.

Azure AD MFA requires at least two of the following authentication methods:

- Something you know—a username and password is the most common way a user would provide credentials.

- Something you have—this could be a trusted device such as your phone or a hardware key.

- Something you are—such as a fingerprint or face scan.

As an administrator, it is possible to define what forms of secondary authentication can be used.

Azure AD Identity Protection

Implementing a user risk security policy using Azure AD Identity Protection can help you to identify potential risks and take manual actions for each of them if needed. There are different risk detection types that you can identify and, as an administrator, you can use key reports for your investigations. Azure AD Premium P2 edition or an EMS E5 subscription is needed to use Azure AD Identity Protection.

There are primarily two risk policies that can be enabled in Azure AD: **sign-in risk policy** and **user risk policy**. While both policies work to automate the response to risk detection in your environment, it is worth understanding the difference between them:

- A sign-in risk policy is an automated response that you can configure for a specific sign-in risk level and block access to resources, requiring the user to pass MFA to prove their identity in order to gain access to a specific resource. When configuring a sign-in risk policy, you need to set the users and groups that it applies to.

- On the other hand, an IT administrator can customize the login experience based on the user's previous risky logins with a user risk policy. A user risk policy is also an automated response that remediates users when they meet a specific risk level or there's a high probability that their credentials have been compromised. This policy will allow you to block access to the resources or require the user to reset their password to keep their identity safe.

Here's a reference for you to configure risk policies: https://docs.microsoft.com/azure/active-directory/identity-protection/howto-identity-protection-configure-risk-policies.

Privileged Identity Management

Privileged Identity Management (**PIM**) is useful for those organizations who are moving to (if not already in) the cloud and are looking to minimize risks, particularly when it comes to privileged accounts assigned to roles in the organization. We are referring to accounts that can access a lot of resources and data.

PIM is a service in Azure AD that helps you reduce these risks by enforcing **Just in Time** (**JIT**) and **Just Enough Access** (**JEA**) for these accounts.

PIM forces administrators in the organization who need to use administrative privileges to elevate their privilege in order to use it, so it requires approval to activate privileged rules and enforces MFA.

You can follow these steps to configure PIM for your organization: https://docs.microsoft.com/azure/active-directory/privileged-identity-management/pim-security-wizard.

Azure AD Application Proxy

The security experience in Azure AD provides you with a one-stop shop for all identity security capabilities. By using Azure AD Application Proxy, you can enable remote work access to on-premises resources. This means that Azure AD Application Proxy allows you to publish on-premises applications to users outside the corporate network, while extending Azure AD capabilities as a central management point for all the applications in the organization.

This brings some benefits, such as MFA and Conditional Access. More importantly, it avoids the limitations of traditional solutions such as VPNs, as there's no need to open access to the entire network. Azure AD Application Proxy allows you to control what resources should be accessible and works across many devices, including mobile and desktop.

An organization can have a corporate network with multiple applications across departments within the DMZ with multiple segments, as well as users outside the network who need to get access to these resources in the corporate network.

In order to enable Azure AD Application Proxy, you first have to install local agents in the network, which are called **connectors**.

These connectors are agents that are executed on a Windows Server machine in your corporate network and are responsible for the communication between your on-premises application and Azure AD Application Proxy.

The connectors can also be integrated with the on-premises Active Directory depending on the single sign-on mechanism that you need. It is important to note that Azure AD Application Proxy requires the Azure AD Premium P1 or P2 license:

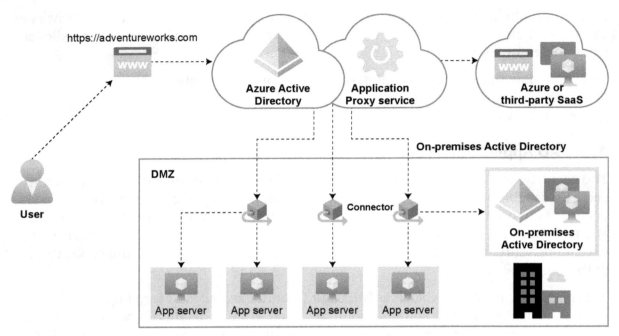

Figure 5.14: Azure AD Application Proxy workflow

The Azure AD Application Proxy workflow, as seen in *Figure* 5.14, is as follows:

1. Your organization can have a public-facing application through an endpoint that can be a URL.

2. External users can connect to the application using that URL.

3. External users will be redirected to a cloud service; the identity provider, which is Azure AD, will authenticate the user using MFA, Conditional Access, or any other method.

4. Once authenticated, Azure AD will send the token to the user's client device.

5. Then, Azure AD will send the token to the Azure AD Application Proxy service, and the user will be re-directed to the on-premises application using the Application Proxy connector.

6. The Azure AD Application Proxy connector will perform additional authentication on behalf of the user if single sign-on is configured.

7. Lastly, the connector sends the request to the on-premises application, and the response is sent back to the user through the connector and Azure AD Application Proxy service.

As you can see, Azure AD Application Proxy can be used for your on-premises applications that need to be accessed from outside the corporate network.

Azure AD Conditional Access

To enable remote work for users trying to reach applications both on-premises and in the cloud, administrators need to ensure the right people get the right permissions to access the resources. Conditional Access is a premium feature in Azure AD that addresses these challenges by providing a mechanism to control access to the resources based on policies.

Using Azure AD Conditional Access, you as an administrator can easily protect your organization's assets and apply the right access controls through policies that can be configured based on some common signals, such as device, application, IP location, and user or group membership.

Examples of Azure AD Conditional Access include:

- **Location**: Users accessing a resource when they're off the corporate network should be required to use MFA.

- **Device platform**: Define a policy for each device platform that blocks access. For instance, allow only iOS devices but not Android.

- **Cloud apps**: Requires that apps which use sensitive information are blocked if Azure AD detects a risky sign-in.

This ends our detailed look at Azure AD. Let's move on to the other ways Azure can help you to enable remote connectivity.

Enabling remote connectivity

When it comes to connectivity, Azure Virtual Network is the fundamental building block to extend your connectivity to Azure and provide the best application experience and connectivity across the organization, partners, and customers. Azure provides a variety of networking capabilities to provide connectivity and enable remote work in a secure manner.

These networking capabilities include several Azure networking services. There are three main services that enable remote work: Azure Virtual WAN, Azure VPN, and ExpressRoute.

Azure Virtual WAN

Azure Virtual WAN provides a unified framework for network security and routing capabilities for you to connect and span across different regions. It also allows you to deploy Azure **Network Virtual Appliances (NVAs)** inside a Virtual WAN hub.

A Virtual WAN hub is basically a virtual network where you can deploy gateways integrated with ExpressRoute and NVAs. In order to enable remote work, you can use Azure Virtual WAN's capabilities to provide a better connectivity experience for remote users. It can connect multiple branches through Site-to-Site VPNs and remote users through Point-to-Site VPNs.

When you use Azure Virtual WAN, you can expand the connectivity with multiple hubs to get a fully meshed hub that allows you to achieve an any-to-any connectivity approach. This way, you can connect a remote user to Azure while at the same time interconnecting an ExpressRoute endpoint or a branch behind a Site-to-Site VPN.

It is important to highlight that every Azure Virtual WAN has a router that controls the routes with all the other gateways in the hub. It also enables network-to-network transit capabilities with up to 50 Gbps in aggregation capacity.

Azure Virtual WAN comes with built-in security capabilities. You can deploy a hub per region and provision multiple firewalls, which can be managed through Azure Firewall Manager. This way, you can use Azure Firewall Manager to work across subscriptions, regions, and deployments, and create policies to secure all the traffic going through the network.

Most organizations today are relying on partners and vendors to enable these networking capabilities through hardware or virtual-based appliances. These can be migrated to Azure as you have the ability to deploy an NVA through the Azure Marketplace and integrate it inside the Virtual WAN hub. Also, there are BGP capabilities built in to enable transit routing across multiple networks and exchange routes that inform on-premises devices and Azure VPN gateways about whether they are reachable or not:

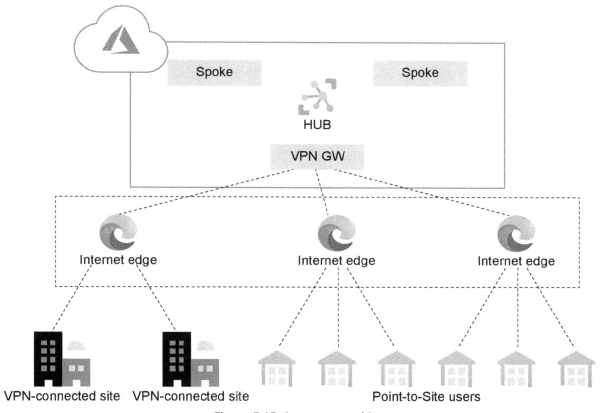

Figure 5.15: Azure networking

You can provision multiple virtual networks in Azure across multiple regions, providing remote workers with a Point-to-Site VPN connection or a Site-to-Site connection for your VPN gateway connection from your on-premises network to the virtual network. This is shown in *Figure 5.15*.

Azure VPN

An Azure VPN gateway enables you to send encrypted traffic across on-premises locations and Azure virtual networks over the internet or between other virtual networks. It is possible to have multiple connections to the same VPN gateway; however, you can only have a single VPN connection per virtual network.

Before provisioning a specific type of VPN gateway, we need to understand the configurations available for VPN gateway connections and evaluate which one is the best fit for your organization. Azure provides multiple connection topologies that you can choose from to meet your organization's needs: **Site-to-Site (S2S)**, multi-site, **Point-to-Site (P2S)**, and ExpressRoute.

Let's discuss each of them in turn.

Site-to-Site and Multi-Site

Site-to-Site VPN gateway is the best fit for your organization if you're looking to achieve hybrid configurations between your on-premises location and virtual networks. S2S is a connection over an IPsec/IKE (IKEv1 or IKEv2) VPN tunnel, which means that in order to deploy this type of connection you will need a VPN device located in your on-premises datacenter and a public IP address assigned to it. The following figure shows an S2S VPN connectivity schema:

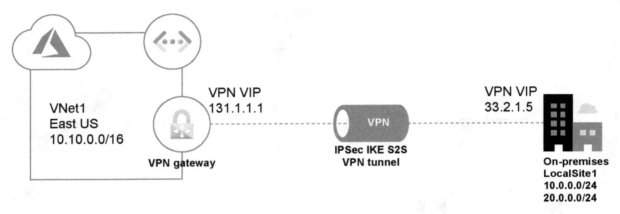

Figure 5.16: Site-to-Site VPN

If your organization has multiple on-premises locations, it is possible to connect branches to the same virtual network using a multi-site connection approach through a route-based VPN. Bear in mind that virtual networks can only have a single VPN gateway; all the connections going through the same VPN gateway will share the available bandwidth:

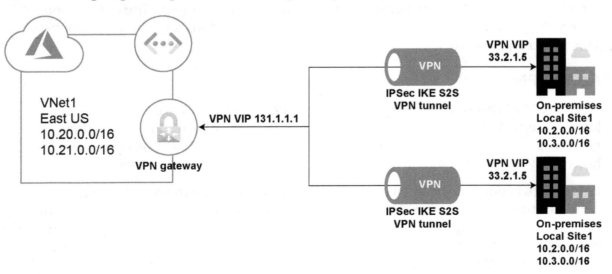

Figure 5.17: Multi-site VPN

In addition, you can configure a network-to-network connection and combine it with a multi-site configuration as seen in *Figure* 5.17. You can establish connectivity with multiple virtual networks in the same or different subscriptions, regions, and deployment models.

Point-to-Site VPN

If you're looking to give remote users access to resources in Azure, you can utilize P2S VPN gateway connections. In this deployment mode, there's no need for a VPN device on-premises; instead, the connection is directly established from the client device as seen in *Figure* 5.18:

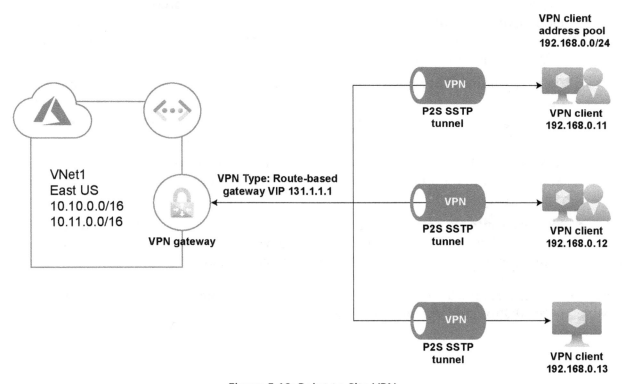

Figure 5.18: Point-to-Site VPN

This type of connection is cost-effective if you want to allow a few remote users to access resources in the cloud through a virtual network.

ExpressRoute

As is very often heard in the field, ExpressRoute is your **Ethernet cable to Azure**. It provides connectivity between your on-premises location and Azure using a private connection through a partner known as the **connection provider**. ExpressRoute offers Layer 3 connectivity, which supports any-to-any network connections, point-to-point Ethernet connections, or a virtual cross-connection.

Based on your organization's needs and the connectivity provider, you can choose how you want to create a connection to the cloud. There are four ways to create a connection:

- **CloudExchange co-location**: If your datacenter is co-located in a facility with a cloud exchange, you can leverage Layer 2 cross-connections, or managed Layer 3 cross-connections through the provider's Ethernet exchange to connect your infrastructure and Azure.

- **Point-to-point Ethernet connection**: On this model, you can connect your datacenter to Azure through Ethernet providers that offer Layer 2 connections, or managed Layer 3 connections and build your private connection using point-to-point Ethernet links.

- **Any-to-Any (IPVPN) connection**: You can enable WAN connectivity between Azure and branch offices or datacenters, usually offered as a managed Layer 3 connectivity.

- **ExpressRoute Direct**: You can plug into Microsoft's global network and achieve an Active/Active connectivity at scale to cover different needs such as massive data ingestion into Azure services or connectivity isolation.

ExpressRoute circuits, as in *Figure* 5.19, provide a wide range of bandwidth, from 50 Mbps up to 10Gbps, and dynamic scaling without downtime:

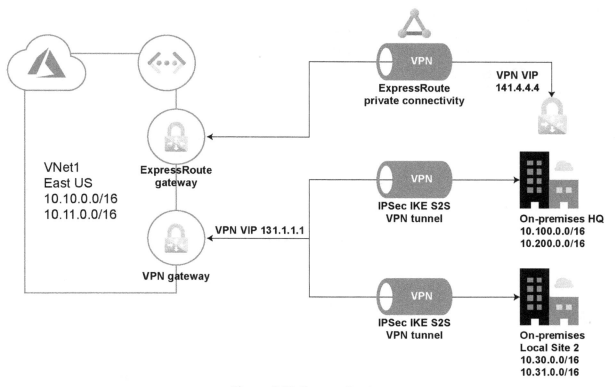

Figure 5.19: ExpressRoute

In addition, it is possible to have an ExpressRoute circuit and an S2S connection at the same time. Although they can coexist, it requires two virtual network gateways for the same virtual network, one using the gateway type VPN, and the other using the gateway type ExpressRoute.

With your connections to Azure established, let's look at how you can secure and manage the traffic to your cloud-based applications.

Securing and managing traffic to your workloads

Security is a shared responsibility between Microsoft, partners, and customers. Azure provides security controls and services to protect identity, network, and data resources.

In the previous section *Azure Active Directory*, we reviewed how you can leverage identity and management features available with Azure AD to ensure secure access to data and resources. Protecting the infrastructure and workloads for remote access on a daily basis is a critical task and Microsoft provides enterprise-grade solutions to secure the traffic to your applications, whether they are delivered through Azure or from an on-premises location.

Azure can help protect your cloud infrastructure by using network security controls to manage traffic, configuring access rules, extending connectivity to your on-premises resources, and protecting your virtual network.

One of the principal services that you can use to protect web applications is **Azure Web Application Firewall (WAF)**, shown in *Figure 5.20*:

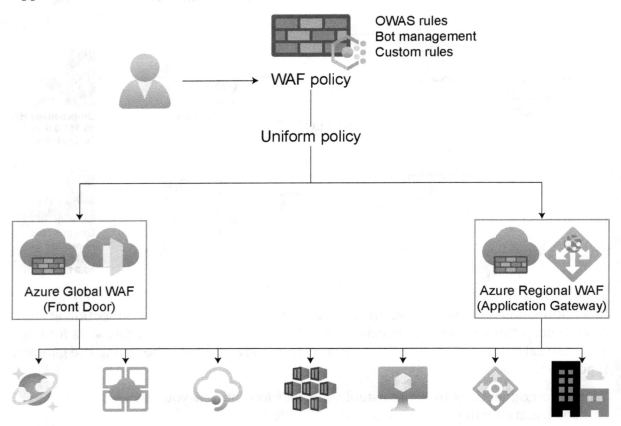

Figure 5.20: Azure WAF

Azure WAF is compliant with various compliance standards, including but not limited to PCI-DSS, HIPAA, SOC, ISO, and CDSA. A complete list of compliance offerings can be found here: https://docs.microsoft.com/azure/compliance/.

Application Gateway with WAF

Your organization may be running applications in the cloud or on-premises, with remote users trying to reach those workloads. Application Gateway is a key **Platform as a Service (PaaS)** component that can facilitate this. It is a lightweight **Application Delivery Controller (ADC)** that provides support for SSL termination, session affinity, and content-based routing. This set of capabilities makes Application Gateway a flexible service. It can be deployed, along with Azure WAF or Azure Front Door, to improve scaling capabilities and protect your applications on-premises or in the cloud against OWASP and other vulnerabilities, without the need for additional configurations.

Both Application Gateway and Azure WAF are PaaS managed services, so you can save some time and provide the right protection for your applications while Azure takes care of the underlying infrastructure, as shown in the following figure:

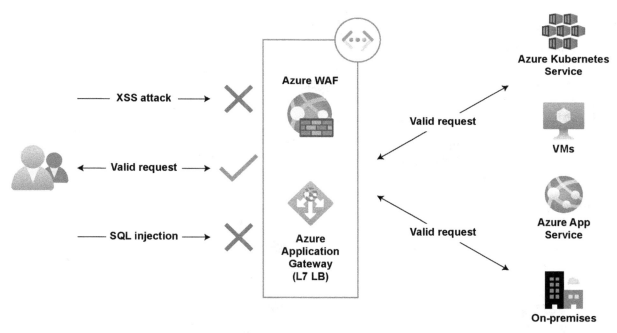

Figure 5.21: Azure Application Gateway and Azure WAF

The key benefits of Azure WAF include:

- Preconfigured protection against the OWASP Top 10 attacks.
- Bot protection.
- A custom rules engine (geo-filtering, IP restriction, HTTP parameter filtering, and size restriction).
- Logs and metrics for detection and alerts.

Another common use case is integration with Azure Front Door, which provides a global application acceleration and delivery service in Azure, in addition to regional load balancing for web applications.

A key benefit of using Azure Front Door with Azure WAF is the ability to detect and prevent attacks against your web applications before they reach the virtual network. You can stop malicious attacks close to the sources before they enter your virtual network and create rate limits or request thresholds to protect your application against floods. We can see how this works in *Figure 5.22*:

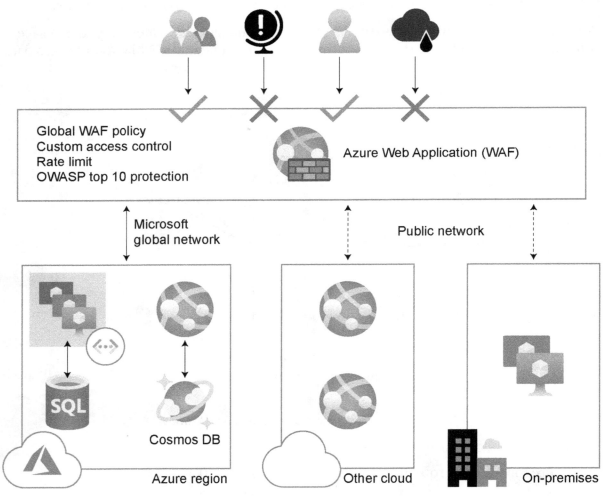

Figure 5.22: Azure Front Door and Azure WAF

This can allow you to publish and secure your web applications whether they are hosted on-premises or in the cloud.

Azure Firewall

The next security control at the perimeter that your organization can use to protect your virtual network resources is Azure Firewall. Azure Firewall allows you to create, enforce, and log applications and network connectivity policies from Layer 3 up to Layer 7 of the OSI model and span across subscriptions and virtual networks.

Azure Firewall is critical when considering remote work as it can provide protection for your Windows Virtual Desktop infrastructure deployments on Azure, along with threat intelligence so remote users can properly access their virtual desktops. It is recommended that you use **Fully Qualified Domain Names (FQDNs)** when using Windows Virtual Desktop to simplify access. Azure Firewall provides a simplified integration for Windows Virtual Desktop by enabling filtering capabilities for outbound traffic, as shown in *Figure 5.23*:

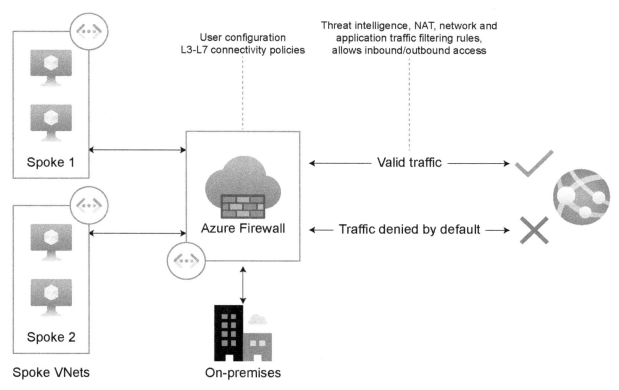

Figure 5.23: Azure Firewall

It is important to note that Azure Firewall is a fully cloud-based managed service and Azure takes care of the underlying infrastructure.

Azure Load Balancer and cross-region load balancing

Very often, organizations struggle with cloud-based workloads when there's an increase in application traffic and want to provide a better customer experience. Azure provides built-in load balancing capabilities for cloud resources so you can create highly available and scalable web applications.

The main component for achieving high availability is Azure Load Balancer, as seen in *Figure 5.24*. It operates at Layer 4 of the OSI model and represents a single point of contact for clients. Azure Load Balancer is optimized for cloud workloads to distribute inbound traffic to back-end pool instances.

Azure Load Balancer can be public facing, so it can load balance internet traffic to your applications living in Azure, or it can be internally facing, where private IPs are needed on the front-end side only:

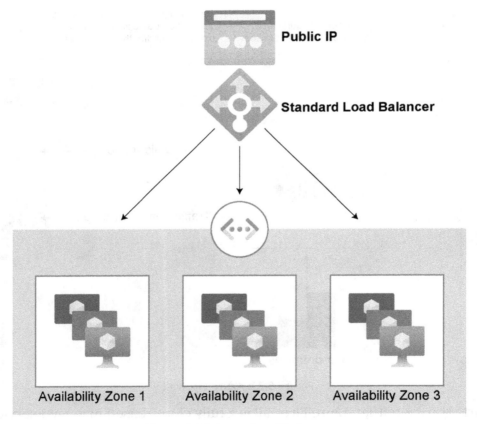

Figure 5.24: Azure Load Balancer

You will find two main Azure Load Balancer profiles, Basic and Standard. Previously, you were only able to add network interfaces associated to virtual machines or virtual machine scale sets in the back-end pool of a load balancer. However, due to recent improvements that allow load balancing across IP addresses, you are now able to load balance resources living in Azure via private IPv4 or IPv6 addresses through the Standard profile, which extends the capability for load balancing across containers.

In the past, Azure Load Balancer was able to load balance only in a given region. Now, you can use Azure Load Balancer for cross-region load balancing to distribute inbound traffic and enable geo-redundant high-availability scenarios:

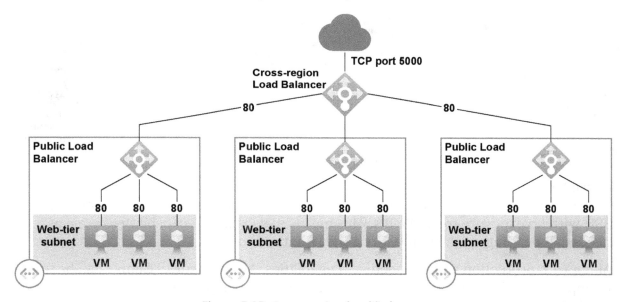

Figure 5.25: Cross-region load balancer

It is possible to build cross-regional capabilities on top of an existing load balancer solution, and a huge benefit is the ability to use a static front-end IP configuration.

Azure Bastion and Just-in-Time (JIT) access

Sometimes organizations need to provide remote access to resources in Azure and want to provide a secure way to connect to these resources without the need of having a VPN connection or assigning public IPs to a virtual machine.

Azure Bastion enables you to manage your resources in a secure manner by providing RDP/SSH connectivity to virtual machines deployed within a virtual network, without needing public IPs directly from the Azure portal:

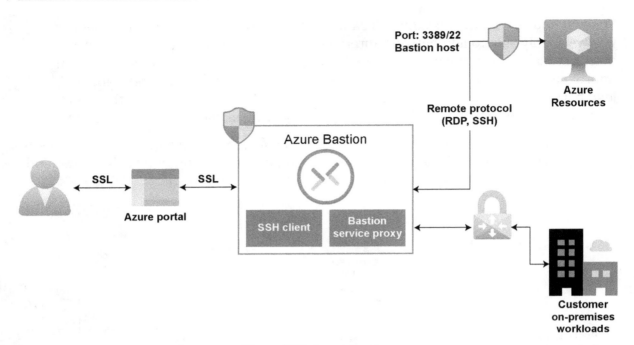

Figure 5.26: Azure Bastion

Azure Bastion is a managed solution and, in the back-end, a virtual machine scale set that is deployed within your virtual network so that remote users can connect via RDP/SSH with no exposure to public IPs. This means it also has the capability to scale as needed depending on the number of concurrent sessions.

Along with protection and connectivity capabilities, Azure Security Center can be used to implement JIT access, which enables you to lock down inbound traffic to your virtual machines to reduce exposure to attacks and easily audit access to the virtual machine.

JIT access can also be used with Azure Firewall to reduce exposure to network volumetric attacks by creating policies to define the ports that you want to protect, for instance, to set how long they should remain open for, and filtering the IPs that should have access to the virtual machine.

Now that we have discussed the main services that can enable your organization to better secure and manage traffic to your applications, let's take a look at the final piece: enabling remote work through the Azure Windows Virtual Desktop infrastructure.

Enabling remote work with Windows Virtual Desktop

In the past few months, we all have experienced an impact on how we collaborate with colleagues within our organizations, as well as partners and customers. Companies of all sizes have invested in the adoption of collaboration tools that can enable remote work and provide the best experience for employees to keep connected during business disruption and improve productivity.

Virtual Desktop Infrastructure (**VDI**) is a virtualization solution that provides a desktop environment for remote users as if they were working on their local computer. Windows Virtual Desktop makes remote work possible while providing the best experience to maintain the underlying infrastructure, and it can even be built in minutes.

Windows Virtual Desktop supports a variety of operating systems, including Windows 10 Enterprise, Windows 10 Enterprise multi-session, Windows 7 Enterprise, and Windows Server 2012 R2, 2016, and 2019. Depending on your organization's needs, you can make use of a specific licensing model such as Microsoft 365 E3, E5, A3, A5, or F3; Business Premium; Windows E3, E5, A3, or A5; or RDS client access licenses. *Figure* 5.27 demonstrates how Windows Virtual Desktop works:

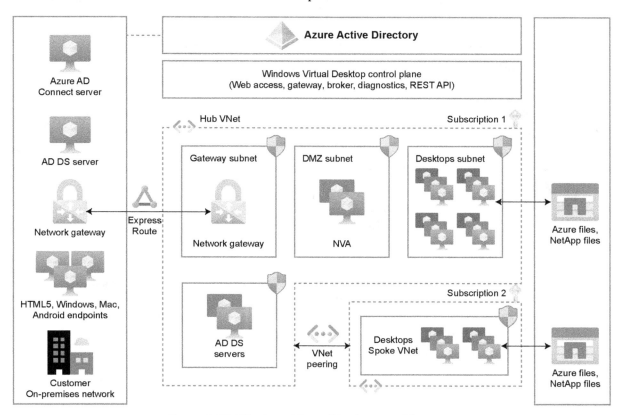

Figure 5.27: Windows Virtual Desktop architecture

- On the left, the endpoints are usually in the on-premises network and can extend connectivity through S2S or Express Route, and Azure AD Connect can integrate the on-premises AD DS with Azure AD.

- Your organization can manage AD DS, Azure AD, Azure subscriptions, virtual networks, storage, and the Windows Virtual Desktop host pools and workspaces.

- Microsoft manages the Windows Virtual Desktop control plane, which handles web access, gateway, broker, diagnostics, and extensibility components such as REST APIs.

In order for your organization to utilize Windows Virtual Desktop, you need Azure AD, a Windows Server Active Directory instance in sync with Azure AD, and an Azure subscription. The virtual machines for Windows Virtual Desktop can be standard domain-joined or hybrid AD-joined.

The new ARM-based experience for Windows Virtual Desktop makes it easier to manage your environment. It includes a new integrated monitoring UI with Log Analytics and management UI through the Azure portal. It also enables you to manage your environment using the Azure PowerShell module for Windows Virtual Desktop.

The ARM-based experience also provides better admin access control as it leverages Azure **Role-Based Access Control** (**RBAC**) with built-in custom roles. You can also publish applications and manage access through Azure AD user groups and individual users. In addition, you are now able to specify the geography you prefer for storage.

The following diagram provides you with a high-level overview of the Windows Virtual Desktop architecture:

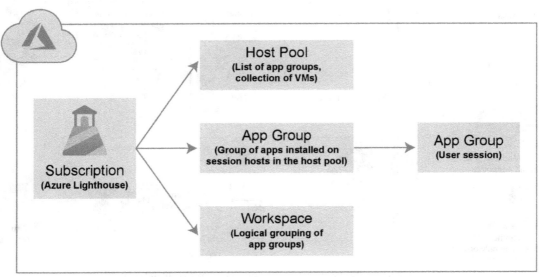

Figure 5.28: Windows Virtual Desktop structure

Through the Azure portal, you will have a seamless experience managing your Windows Virtual Desktop deployment, including host pools, application groups, workspaces, and users:

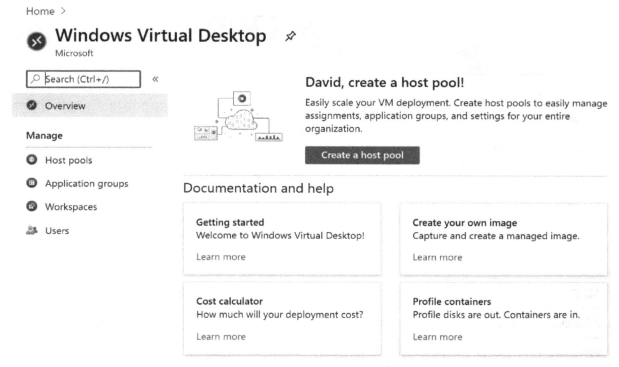

Figure 5.29: Managing Windows Virtual Desktop through the Azure portal

You can use ARM capabilities to provision Windows Virtual Desktop. You need to register the `Microsoft.DesktopVirtualization` resource provider, then create the host pools as needed, including the specification of the virtual machine details, and provide the parameters for the Azure virtual network. After this, you can proceed to create the application groups and add Azure AD users or user groups to them. Finally, you configure your Windows Virtual Desktop workspace and then install the Windows Virtual Desktop client for Windows on the client device.

There's a learning path on Microsoft Learn on how to deliver remote desktops and apps from Azure with Windows Virtual Desktop: https://docs.microsoft.com/learn/paths/m365-wvd/.

Also, you can leverage the ARM templates to automate the deployment process of your Windows Virtual Desktop environment: https://github.com/Azure/RDS-Templates/.

Components you can manage in Windows Virtual Desktop

The following table compares the resources that Microsoft manages in Windows Virtual Desktop and the ones that you can take care of:

Component	Managed by	
	Microsoft	Customers
Web access	✔	
Gateway	✔	
Connection broker	✔	
Diagnostics	✔	
Extensibility components	✔	
Azure virtual network		✔
Azure AD		✔
AD DS		✔
Windows Virtual Desktop session hosts		✔
Windows Virtual Desktop workspace		✔

Table 5.2: Responsibilities for the management of resources

These are the components managed by Microsoft:

- **Web access**: Through this service, users are able to access their desktops using an HTML5-compatible web browser.

- **Gateway**: This service enables remote users to establish a connection with Windows Virtual Desktop apps and desktops from any device that can execute the Windows Virtual Desktop client.

- **Connection broker**: This service manages user connections and provides load balancing capabilities.

- **Diagnostics**: This is an event-based aggregator to identify potential failures of the components.

- **Extensibility components**: This includes third-party tools and additional extensions that can be enabled for Windows Virtual Desktop.

These are the components you manage:

- **Azure virtual networks**: You can configure the network topology as needed to provide access to desktops and applications based on your organization's policies.

- **Azure AD**: While this is a pre-requisite for using Windows Virtual Desktop, you can manage Azure AD integration and security features to keep your environment compliant.

- **AD DS**: This is a pre-requisite for enabling Windows Virtual Desktop that includes synchronization with Azure AD. You can also leverage Azure AD Connect to associate AD DS with Azure AD.

- **Windows Virtual Desktop session hosts**: You can run a variety of operating systems, including Windows 10 Enterprise; Windows 10 Enterprise multi-session; Windows 7 Enterprise; Windows Server 2012 R2, 2016, and 2019; and also custom Windows images.

- **Windows Virtual Desktop workspace**: This is a workspace for managing and publishing host pool resources.

Sizing your Windows Virtual Desktop environment

In the past, sizing your Windows Virtual Desktop environment was a bit of a challenge due to the availability of only some basic guidance. Now you can utilize the Windows Virtual Desktop Experience Estimator, a tool that can help you to find the right size for your environment for your organization's needs.

The Windows Virtual Desktop Experience Estimator is available at https://azure.microsoft.com/services/virtual-desktop/assessment/.

Initially, you will find a table that shows the region with the lowest round trip time from your current location and also provides a reference to the size of the virtual machines and network services needed for your implementation:

Azure Region*	Round Trip Time (ms)
East US 2	15
East US	20
North Central US	41
South Central US	41
Canada Central	43
Central US	47
Canada East	54
West Central US	62

Table 5.3: Round trip times for different Azure regions

Based on your current location, you can choose a preferred region to provide a better user experience. Bear in mind that there are also additional components that might impact your deployment, such as network conditions, end user devices, and the configuration of the virtual machines.

Network guidelines

As we have already discussed, networking components are critical to enabling remote work. Bandwidth is one of the key metrics that you should look at when sizing your environment as it will impact the user experience. The following table demonstrates some recommendations based on different types of workloads and can be found at https://docs.microsoft.com/windows-server/remote/remote-desktop-services/network-guidance:

Workload type	Recommended bandwidth
Light	1.5 Mbps
Medium	3 Mbps
Heavy	5 Mbps
Power	15 Mbps

Table 5.4: Recommended bandwidth for different workload types

The display resolution is another form factor along with the bandwidth. The following table shows the recommended bandwidth for an optimal user experience:

Typical display resolutions at 30 fps	Recommended bandwidth
About 1024 × 768 px	1.5 Mbps
About 1280 × 720 px	3 Mbps
About 1920 × 1080 px	5 Mbps
About 3840 × 2160 px (4K)	15 Mbps

Table 5.5: Recommended bandwidth for different display resolutions at 30 fps

With regard to the virtual machines for your Windows Virtual Desktop environment, Microsoft provides examples of different workload types to properly size the virtual machines that you need for your environment:

Workload type	Example users	Example apps
Light	Users doing basic data entry tasks	Database entry applications, command-line interfaces
Medium	Consultants and market researchers	Database entry applications, command-line interfaces, Microsoft Word, static web pages
Heavy	Software engineers, content creators	Database entry applications, command-line interfaces, Microsoft Word, static web pages, Microsoft Outlook, Microsoft PowerPoint, dynamic web pages
Power	Graphic designers, 3D model makers, machine learning researchers	Database entry applications, command-line interfaces, Microsoft Word, static web pages, Microsoft Outlook, Microsoft PowerPoint, dynamic web pages, Adobe Photoshop, Adobe Illustrator, **computer-aided design (CAD)**, **computer-aided manufacturing (CAM)**

Table 5.6: Examples of different workload types

In addition, an example of a metric that you should take into consideration based on the workloads is the number of users reaching the same application at the same time. That said, you can consider the multi-session or single-session recommendations based on your organization's needs.

Microsoft provides a virtual machine size based on the maximum suggested number of users per **virtual CPU (vCPU)** and examples of Azure instances that match the minimum virtual machine configuration:

Workload type	Maximum users per vCPU	vCPU/RAM/OS storage minimum	Example Azure instances	Profile container storage minimum
Light	6	2 vCPUs, 8 GB RAM, 16 GB storage	D2s_v3, F2s_v2	30 GB
Medium	4	4 vCPUs, 16 GB RAM, 32 GB storage	D4s_v3, F4s_v2	30 GB
Heavy	2	4 vCPUs, 16 GB RAM, 32 GB storage	D4s_v3, F4s_v2	30 GB
Power	1	6 vCPUs, 56 GB RAM, 340 GB storage	D4s_v3, F4s_v2, NV6	30 GB

Table 5.7: Multi-session recommendations

As a general recommendation, it is a good idea to use Premium SSD disks for critical workloads and consider GPUs if you plan to run graphics-intensive programs on your virtual machines.

Managing user profiles with Azure Files and NetApp Files

While Azure offers a wide range of storage solutions, FSLogix profile containers are one of the recommended user profile solutions for a Windows Virtual Desktop service as they are designed to roam profiles in non-persistent remote computing environments.

When a user performs a sign-in, a container is dynamically attached to the computing environment. You can also create profile containers using Azure NetApp Files to quickly provision SMB volumes for Windows Virtual Desktop.

The following table provides a comparison between the different solutions available for Windows Virtual Desktop FSLogix profile container user profiles:

Features	Azure Files	Azure NetApp Files	Storage Spaces Direct
Use case	General purpose	Ultra-performance or migration from NetApp on-premises	Cross-platform
Platform service	Yes, Azure-native solution	Yes, Azure-native solution	No, self-managed
Regional availability	All regions	Select regions	All regions
Redundancy	Locally redundant/ zone-redundant/ geo-redundant	Locally redundant	Locally redundant/ zone-redundant/geo-redundant
Tiers and performance	Standard Premium Up to max 100k IOPS per share with 5 GBps per share at about 3 ms latency	Standard Premium Ultra Up to 320k (16K) IOPS with 4.5 GBps per volume at about 1 ms latency	Standard HDD: up to 500 IOPS per-disk limits Standard SSD: up to 4k IOPS per-disk limits Premium SSD: up to 20k IOPS per-disk limits We recommend Premium disks for Storage Spaces Direct
Capacity	100 TiB per share	100 TiB per volume, up to 12.5 PiB per subscription	Maximum 32 TiB per disk
Required infrastructure	Minimum share size 1 GiB	Minimum capacity pool 4 TiB, min volume size 100 GiB	Two virtual machines on Azure IaaS (+ Cloud Witness) or at least three virtual machines without and costs for disks
Protocols	SMB 2.1/3. and REST	NFSv3, NFSv4.1 (preview), SMB 3.x/2.x	NFSv3, NFSv4.1, SMB 3.1

Table 5.8: Windows Virtual Desktop FSLogix profile container user profiles

To look up the regional availability details for Azure NetApp Files and other products, visit https://azure.microsoft.com/global-infrastructure/services/.

The following table provides a comparison of the storage solutions Azure Storage offers for Windows Virtual Desktop that are supported on Azure with FSLogix profile container user profiles:

Features	Azure Files	Azure NetApp Files	Storage Spaces Direct
Access	Cloud, on-premises, and **hybrid (Azure file sync)**	Cloud, on-premises (via ExpressRoute)	Cloud, on-premises
Backup	Azure backup snapshot integration	Azure NetApp Files snapshots	Azure backup snapshot integration
Security and compliance	All Azure supported **certificates1**	ISO completed	All Azure supported **certificates1**
Azure Active Directory integration	Native Active Directory and Azure AD DS2	Azure AD DS and Native Active Directory3	Native Active Directory or Azure AD DS support only

Table 5.9: Storage solutions Azure Storage offers for Windows Virtual Desktop

Here are some relevant links to further information:

1. Microsoft compliance offerings: https://docs.microsoft.com/compliance/regulatory/offering-home

2. Azure Files integration with AD: https://docs.microsoft.com/azure/storage/files/storage-files-active-directory-overview

3. Azure NetApp Files integration with AD: https://docs.microsoft.com/azure/azure-netapp-files/azure-netapp-files-faqs#does-azure-netapp-files-support-azure-active-directory

Azure Monitor for Windows Virtual Desktop

It is important for you to minimize the risk of failures in the Windows Virtual Desktop environment. Azure provides native integration with Azure Monitor to quickly identify issues through a single dashboard.

You can review the following activity logs:

- Management activities
- Connections
- Host registration
- Errors
- Checkpoints

In order to monitor your Windows Virtual Desktop environment, you need to create a Log Analytics workspace via the Azure portal or PowerShell and then connect the virtual machines to Azure Monitor.

Once you have onboarded your Windows Virtual Desktop into Log Analytics, you will be able to run custom queries using **Kusto Query Language** (**KQL**) to analyze the health of your Windows Virtual Desktop environment. A simple use case could be if there's a need to find session duration by the user, as shown in the following code:

```
let Events = WVDConnections | where UserName == "userupn" ;
Events
| where State == "Connected"
| project CorrelationId , UserName, ResourceAlias , StartTime=TimeGenerated
| join (Events
| where State == "Completed"
| project EndTime=TimeGenerated, CorrelationId)
on CorrelationId
| project Duration = EndTime - StartTime, ResourceAlias
| sort by Duration asc
```

In this code, we are using a `let` statement for the `Event` instance called `WVDConnections` and filtering tables for rows that match the users with a connected state. Then, as good practice, we use `project` to select just the columns we need before performing the `join`. We pull events with the first `EventType` and with the second `EventType` and then join the two sets on `CorrelationId`.

After this, we rename the timestamp column and sort the results in ascending order.

You can use Azure Monitor workbooks to create custom visualizations of your Windows Virtual Desktop performance. You can find an example of a workbook for Windows Virtual Desktop here: https://github.com/wvdcommunity/AzureMonitor/blob/master/WVD-ARM-monitoring-workbook.json.

Windows Virtual Desktop partner integrations

Microsoft partners are a key component in order to fully enable an organization's remote work capabilities; there are Windows Virtual Desktop partner integrations with Citrix and VMware.

Citrix extended its Virtual Apps and Desktops offering to Azure, which can be consumed as a managed service and is available through the Azure Marketplace. This is a combination of new technology and new entitlements. It supports a multi-session Windows 10 experience and, if you have the appropriate RDS license, you can use Windows Server. It also supports persistent and non-persistent single-session or multi-session experiences.

You can read more about Virtual Apps and Desktops for Azure here: https://azure.microsoft.com/services/virtual-desktop/citrix-virtual-apps-desktops-for-azure/.

VMware is an approved Windows Virtual Desktop provider that simplifies the deployment of hybrid Windows Virtual Desktop environments through its Horizon Control Plane to enable a single management interface for both on-premises and cloud-based deployments.

Customers can leverage Windows 10 multi-sessions and the platform is available via the Horizon Universal License, which includes the ability to deploy VMware Horizon on any supported platform:

- You can read more about VMware Horizon Cloud on Microsoft Azure here: https://azuremarketplace.microsoft.com/marketplace/apps/vmware-inc.hc-azure

- Quickstart Guide for Windows Virtual Desktop: https://azure.microsoft.com/resources/quickstart-guide-to-windows-virtual-desktop/

- Windows Virtual Desktop Migration Guide for RDS: https://azure.microsoft.com/resources/windows-virtual-desktop-migration-guide-for-remote-desktop-services/

Summary

In this chapter, we reviewed common infrastructure-related challenges that organizations face when trying to enable remote work, and how Azure AD can help your organization keep its employees' identities and resources protected. We also learned how you can use Azure networking components to enable connectivity and secure access to your organization's workloads both on-premises and in the cloud. Lastly, we reviewed how you can provide a better end user experience through Windows Virtual Desktop and saw guidance on the sizing and configuration of your environment.

The next chapter will provide you with a security baseline and recommendations on how you can use the security services available to your organization so that you can apply standard security control frameworks to your Azure deployments and enforce governance across all your resources.

6

Security fundamentals to help protect against cybercrime

This chapter will help you understand the Azure security fundamentals and guide you through the security services available. We will review how you can use Microsoft services to adopt a Zero Trust strategy in your organization, along with solutions that you can utilize to protect your workloads and assess and improve your security posture. In this chapter, we will cover the following:

- Enabling security for remote workers
- Gaining visibility into your entire infrastructure using **Azure Security Center**
- Achieving advanced threat protection for your hybrid and multicloud environments with **Azure Defender**
- Securing your network
- Modernizing security operations with **Azure Sentinel**
- A unified SecOps experience

Before we drill down into the services you can use to protect your environments, it is important to understand how you can effectively implement security practices to protect your environments whether they are on-premises or in the cloud. So let's begin by addressing the main challenges we face today and how you can start working on an effective security strategy for your organization.

Enabling security for remote workers

COVID-19 has not only impacted businesses of different sizes and verticals but has also fueled cybercrime. The fear related to this pandemic has led cybercriminals with different skillsets and motivations to capitalize by imitating trusted sources and national health organizations using different tactics and techniques.

Enabling remote work has been a challenge, especially for those organizations that were used to providing access to on-premises applications only within their corporate network. Security is a shared responsibility between Microsoft and it's customers; in today's world, a security breach into an organization could mean millions of dollars' worth of damage.

In light of the pandemic, organizations of all sizes have had to adopt new systems and controls to secure access to both on-premises and cloud resources.

Security operations excellence

As many employees are likely to continue working remotely, there's a need to enable hybrid working environments. Using **Zero Trust** principles can help organizations ensure their employees have access to the resources they need, whether those resources are in their private datacenter or the cloud.

A Zero Trust approach involves constant verification and monitoring of access to all corporate services, applications, and network connections. Simply put, "never trust, always verify."

By adopting a Zero Trust security model, organizations can move away from using traditional virtual private networks to access legacy applications by migrating them to Microsoft Azure.

Cloud applications can be accessed through the internet, and for those applications that are not able to fully migrate to Azure. These can be published through **Azure Active Directory (Azure AD)** Application Proxy as we reviewed in *Chapter 5, Enabling secure, remote work with Azure AD and WVD.*

Employees will have the ability to access applications they use on a daily basis through **Windows Virtual Desktop (WVD)** in a more secure way and restrict movement to other resources. This approach gives your organization the ability to align a cloud-first security strategy and ensures employees get access to the applications they need, when they need them, and with the right access levels.

Microsoft provides a security baseline to protect people, devices, infrastructure, and data from vulnerabilities through identity-driven security solutions. Therefore, Microsoft strengthens products and services to identify and address vulnerabilities quickly and efficiently. There are two core platforms available for you to ensure you can adopt best practices in your security strategy:

- Security Engineering Portal
- Microsoft Security Response Center

Microsoft is continually improving its protection of services and data through operational management and threat-mitigation practices. You can use the Security Engineering Portal to learn about the security practices that Microsoft is using to secure applications and services.

On the other hand, you can use **Microsoft Security Response Center (MSRC)** to keep your systems protected and manage security risks. On the MSRC portal, you will see the latest security updates and details on the product family, the severity and impact of vulnerabilities, and get access to download security updates that can be applied to your system.

Cyber Defense Operations Center

Microsoft Cyber Defense Operations Center carries out intrusion detection and responds to compromises and attacks, helping you to better secure a remote workforce. It protects cloud infrastructure and the services, products, and devices that customers use, as well as Microsoft's internal resources, beyond the traditional security perimeter.

Microsoft Cyber Defense Operations Center groups the company's security response experts to help protect, detect, and respond 24x7 to security threats against Microsoft's IT infrastructure and assets globally.

Microsoft's cybersecurity posture is committed to protecting, detecting, and responding to cybersecurity threats. Based on these three pillars, explained in the following sections, Microsoft can provide a useful framework for security strategies and capabilities.

Protect

Microsoft is committed to protecting the compute resources used by customers and employees to ensure resiliency of both the cloud infrastructure and services. Microsoft's protection tactics rely on nine main categories:

- Physical environment monitoring and controls, such as physical access to datacenters, screening, and access approval at the facility's perimeter, inside the building, and on the datacenter floor.
- Software-defined networking components to protect cloud infrastructure from intrusions and attacks such as **distributed denial of service (DDoS)** attacks.
- Identity and management controls with multi-factor authentication to ensure critical resources and data are protected.
- The use of **just-enough-administration (JEA)** and **just-in-time (JIT)** privileges to ensure the right access to resources.
- Configuration management and proper hygiene, maintained through anti-malware software and system updates.
- Identification and development of malware signatures deployed in Microsoft infrastructure for advanced detection and defense.
- Hardening all applications, online services, and products through Microsoft **Security Development Lifecycle (SDL)**.
- Minimizing the attack surface by restricting services.
- Taking appropriate measures to protect classified data, enabling encryption in transit and at rest, and enforcing the least-privilege access principle.

Detect

Although Microsoft continues to invest in these protection layers, cybercriminals are always looking for ways to exploit them. There's no environment that is fully protected, as systems can fail and people can make errors. Therefore, Microsoft has adopted an **Assume Breach** position to ensure that a compromise is rapidly detected, and appropriate actions are taken.

Microsoft's detection tactics rely on six main categories:

- Monitoring network and physical environments
- Highlighting abnormal activity based on identity and behavior analytics
- Machine learning tools to discover irregularities
- Identifying anomalous activities through analytical tools and processes
- Increasing effectiveness by using automated software-based processes
- Determining remediation processes and response to anomalies on systems

Respond

Accelerating triage, mitigation, and recovery are all part of Microsoft's commitment to provide relevant and actionable information through automated response systems. The main objective is to secure vulnerabilities, mitigate attacks, and moreover, respond to cybersecurity events.

Microsoft's response tactics rely on the following six main categories:

- Flagging events that require intervention through automated response systems
- Rapid response by providing a well-defined, documented, and scalable incident response process available to all responders
- Expertise across Microsoft teams to ensure a deep understanding of the platforms, services, and applications operating in cloud datacenters to address incidents
- Determining incident scopes through wide enterprise searching across both cloud and on-premises data and systems
- A better understanding of incidents through deep forensic analysis
- A rapid response time and recovery through security software tools and automation

Organizations can learn from the Microsoft Cyber Defense Operations Center and adopt best practices to secure their environments and improve their security posture, as well as using solutions and integrations available in Microsoft Intelligent Security Association.

Microsoft Intelligent Security Association

Microsoft Intelligent Security Association (**MISA**) is an ecosystem of independent software vendors from across the cybersecurity industry with the ultimate goal of improving an organization's security by sharing their expertise and integrating their solutions, to better defend against a world of increasing threats.

MISA membership benefits include:

- Microsoft security products to extend solution capabilities
- A go-to-market strategy to enjoy co-marketing opportunities
- Customer connections
- Access to product teams to differentiate your solutions

MISA provides guidance and resources for building apps, workflows, and integration with Microsoft security management, threat protection, information protection, and identity and access management solutions so you can build connected security solutions and enable experiences for cross-product scenarios.

You can find members of MISA through the following URL: https://www.microsoft.com/misapartnercatalog

In this section, we have reviewed how Microsoft ensures security operations excellence and tactics that a company should follow to protect, detect, and respond to potential security threats.

In order to improve your security posture, it is important to have visibility across your environment and the ability to use mechanisms to prevent security threats. In the next section, we will go through the key capabilities of Security Center and how you can use this service.

Gain visibility into your entire infrastructure with Azure Security Center

Unified visibility across your environment is critical to monitoring your security posture, detecting threats, and quickly responding to them. Security Center can provide the mechanisms to monitor your environments and proactively respond to security threats.

If you navigate to the Azure portal and search for Security Center, you will be able to see the service on your subscription as shown in *Figure 6.1*:

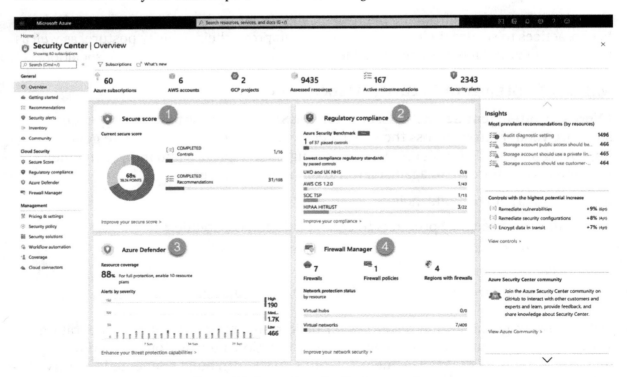

Figure 6.1: Azure Security Center overview

On the **Security Center** page, you can see an overview of the security posture of the resources in your subscription along with recommendations for hardening the security of those resources:

1. **Secure score**: The core of Azure Security Center's cloud security posture management capabilities is the Secure Score. This is a measure of the current status of security recommendations for all the resources. It's calculated using the ratio between your healthy resources and your total resources and a weighting of the importance of some of the security recommendations versus others. The higher the score, the lower the identified risk level.

2. **Regulatory compliance**: Provides insights into the level of your compliance with the various regulatory and industry standards you've applied to your subscriptions. Azure Defender must be enabled to monitor your compliance score.

3. **Azure Defender**: This is Security Center's integrated cloud workload protection platform. Access Azure Defender from within Security Center and protect your hybrid and multicloud workloads running on servers, SQL, storage, containers, and IoT. Defender is part of Microsoft's **extended detection and response (XDR)** solution.

4. **Firewall Manager**: The integration of Azure Firewall Manager into Azure Security Center's main dashboard allows customers to check the firewall coverage status across all networks and to centrally manage Azure Firewall policies.

Security posture management and multicloud

With the rapid adoption of cloud services across multiple cloud providers, organizations are able to streamline operations, enable remote work, and improve employee productivity. The use of cloud applications across multiple cloud platforms introduces new risks and threats.

Cloud Security Posture Management (CSPM) is a term used to address the key attributes that a security solution should include to provide the visibility and capabilities to help you understand your environment, and how you could better secure it, whether it's on the cloud or on-premises.

With Azure Secure Score, Security Center can help you improve your security posture and reduce the likelihood of getting your resources compromised. This extends beyond just IaaS or PaaS, and also applies to SaaS. In fact, *The Total Economic Impact of Azure Security Center Study* points out that Azure Security Center can reduce the risk of a security breach by 25%.

You can read more about the TEI study at the following URL: https://query.prod.cms. rt.microsoft.com/cms/api/am/binary/RWxD0n

Microsoft's cloud security platform can improve the security configuration of your cloud infrastructure, and help you detect and protect your environments in Azure, on-premises, and on other clouds.

Organizations that have adopted a multicloud approach are now able to easily improve their security posture, as Microsoft provides a unified infrastructure security management system through Security Center to strengthen the security posture of both your cloud and on-premises resources while providing threat protection and deep analysis. Through Security Center, you can continuously assess the security state of your resources, including those that are running in other clouds and on-premises datacenters.

Security Center provides you with the tools needed to secure your services and harden your network. You can turn on Security Center in your Azure subscription and use the extended detection and response capabilities to enable threat protection for environments running in Azure, on-premises, and in other clouds.

Most organizations that move to Azure typically migrate applications to the cloud that have some legacy security controls in place to try to protect these workloads. In reality, cloud threats differ from on-premises threats. Therefore, Security Center addresses the need for a cloud-native tool that provides the ability to identify potential vulnerabilities on your resources and provides recommendations on how to address those vulnerabilities.

There are multiple strategies, tactics, and tools to use and secure your hybrid and multicloud environments. Here's a proposed strategy for securing your multicloud environment in four phases:

Discover	Monitor	Access	Protect
Understand your multicloud resources	Detect suspicious activities across workloads	Review misconfigurations and compliance status	Automate protection and policy enforcement of resources in real time

Figure 6.2: Four phases to secure your multicloud environment

Phase 1: Identifying the status of your organization's security posture

It is important to understand and evaluate the status of your resources and their actual usage; to do this, you can use **Cloud Discovery** to analyze all the traffic going through your network. This process of identifying anomalies can be done through machine learning, an anomaly detection engine, or through custom policies that are focused mainly on the discovery and investigation processes.

Phase 2: Detecting suspicious behavior across workloads

It is recommended that you use cloud monitoring tools so you can learn from the alerts, tune activity detections to identify true compromises, and improve the process of handling large volumes of false positive detections. There are some specific recommendations you can follow such as configuring IP address ranges, adjusting the monitoring usage and alerts sensitivity, and tuning anomaly detection policies.

Phase 3: Assessing and remediating misconfigurations and compliance status

Security Center can assess all the resources in a given subscription and provide you with recommendations for the entire environment. When you select a specific recommendation, Security Center will redirect you to the recommendation details page, where you'll see additional details and instructions for how to remediate the identified issue.

For a multicloud approach, if your organization is using **Amazon Web Services (AWS)** or **Google Cloud Platform (GCP)**, it is possible to drill down on the security details on the configuration for AWS. When you connect your AWS to Azure Security Center you will be able to see the AWS recommendations within Security Center, helping you cover a multicloud approach.

Phase 4: Automate protection and policy enforcement for cloud resources

This phase is focused on protecting your cloud resources from data leaks in real time through the enforcement of policies for control and access to resources. This will prevent data exfiltration and the uploading of malicious files to your cloud platforms.

It is possible to create a session policy that blocks the uploading of incorrectly labeled files and configures policies to enforce the correct creation of labels. This way, you can ensure that the data saved to the cloud has the correct access permissions applied.

Azure Security Center architecture

Since Security Center is natively integrated with Azure, you can monitor almost all the resources that are part of your environment and protect them without any additional deployment. As previously mentioned, Security Center can protect non-Azure servers and support Windows and Linux servers through the installation of Security Center's agent, as shown in *Figure 6.3*:

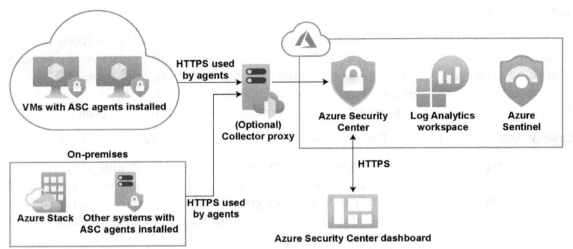

Figure 6.3: Azure Security Center and Azure Sentinel architecture

The agent collects all the security events, which are then correlated in the security analytics engine to provide you with tailored recommendations based on your defined security policies. Security policies are based on policy initiatives created in Azure Policy and can be used to define the desired configuration of your workloads to ensure you're complying with the security requirements of your organization or any specific regulation.

Security Center includes three main security policies:

- **Built-in default policies**: These are policies automatically assigned when you enable Security Center. They are part of a built-in initiative and you can customize them later on.
- **Custom policies**: You can use custom security policies and add your own custom initiatives to receive recommendations if your environment doesn't follow the policies you create. You can configure a custom initiative through Security Center, select the subscription or management group to which you would like to add the custom initiative, and define the parameters for it.
- **Regulatory compliance policies**: You will be able to see a compliance dashboard that shows the status of all the assessments within your environment, to help improve your regulatory compliance.

Azure Security Benchmark

Security is a joint effort between Microsoft and your organization and customers. We have to ensure our workloads remain secure whether they are living on-premises or in the cloud. In order to help you improve the security of your workloads, data, and services running in the cloud, Microsoft provides **security baselines** for Azure to focus on cloud-centric control areas and applies prescriptive guidance and best practices to strengthen security through the **Azure Security Benchmark (ASB)**.

The ASB mainly focuses on cloud-centric control areas. This way you can follow recommendations and guidance that comprise the Cloud Adoption Framework, the Azure Well-Architected Framework, and Microsoft Security Best Practices.

It is important to understand the context and terminology used in Azure security baselines, so we need to emphasize some terms:

- **Control**: This is a description of a feature or activity, not specific to a technology, that needs to be addressed. For example, **Data Protection** is a security control that contains specific actions that should be addressed to ensure your data is protected.
- **Benchmark**: This contains security recommendations for a specific technology and can be categorized by the control it belongs to. For instance, ASB comprises security recommendations specific to Azure.
- **Baseline**: This is the implementation of the security benchmark on Azure services. For example, an organization can enable Azure SQL security features following the Azure SQL security baseline.

If your organization is new to Azure and is looking for recommendations to secure deployments and improve the security posture of existing environments, or trying to meet regulatory requirements for customers who are highly regulated, ASB is the ideal starting point. It includes tooling, tracking, and security controls and is consistent with well-known security benchmarks, such as those described by the **Center for Internet Security (CIS)** Controls Version 7.1 and **National Institute of Standards and Technology (NIST)** SP800-53.

ASB can help you secure the services you use in Azure through a collection of security recommendations that include:

- **Security controls**: These are recommendations that you can apply across your Azure tenant and services.
- **Service baselines**: These can be applied to Azure services to get recommendations on the configuration of the security from a specific service.

To implement ASB, you can start by planning your implementation and validating the enterprise controls and service-specific baselines that might better fit your organization's needs. Next, monitor your compliance using the Security Center's regulatory compliance dashboard. Lastly, establish guardrails to enforce compliance through **Azure Blueprints** and **Azure Policy**. ASB includes the following control domains:

- Network Security
- Identity Management
- Privileged Access
- Data Protection
- Asset Management
- Logging and Threat Detection
- Incident Response
- Posture and Vulnerability Management
- Endpoint Security
- Backup and Recovery
- Governance and Strategy

ASB is the default security policy for Azure Security Center so that you can extend the richness of security recommendations in Azure. As a result, Azure Secure Score will reflect a much broader set of recommendations and spans a broader set of Azure resources.

Additionally, the full control set layout of ASB in the compliance dashboard is now available to all Azure Security Center customers, including Azure Security Center free tier as well as existing Azure Defender customers. Customers can view their compliance relative to the benchmark controls in the compliance view while viewing the detailed impact on their Secure Score. By prioritizing the remediation of security recommendations using Secure Score metrics, customers can achieve a higher Secure Score and attain their compliance goals, all at the same time.

You can review the latest updates of the full Security Center security baseline mapping file in the following GitHub repository: https://github.com/MicrosoftDocs/SecurityBenchmarks/.

Achieve advanced threat protection for your hybrid and multicloud environments with Azure Defender

As the number of home-based workers has accelerated in the last few months, organizations have experienced a far greater need to ensure employees have access through multiple types of devices to company resources. Therefore, it is important to adjust security policies to enable remote work and keep your data secure. As mentioned in the previous section, Security Center can improve your **Cloud Security Posture Management (CSPM)** and provide **Cloud Workload Protection (CWP)** capabilities.

You can use Security Center for free in any given subscription, which includes the detection of security misconfigurations in your Azure resources, a Secure Score to improve your hybrid cloud posture, and enabling Azure Defender for advanced protection for your Azure hybrid workloads.

Azure Defender is integrated into the same cloud-native service as Security Center and provides additional security features such as security alerts and advanced threat protection.

Azure Defender includes native support for Azure services like virtual machines, SQL databases, storage, containers, web applications, networks, and non-Azure servers hosted on-premises or in other clouds such as AWS and GCP.

To extend hybrid cloud protection for your SQL databases and servers running in other clouds or on-premises, it is recommended that you use Azure Arc and Azure Defender. You can find the details on the resource coverage and pricing at the following site: https://azure.microsoft.com/pricing/details/azure-defender/

Azure Defender dashboard

If you navigate to the Azure portal and look for **Security Center**, within **Security Center** you will find the **Azure Defender** dashboard as in *Figure 6.4*:

Figure 6.4: Azure Security Defender dashboard

The **Azure Defender** dashboard will provide you with the overall coverage of your resources, security alerts, advanced protection, and insights into your most attacked resources along with virtual machines' vulnerability alerts:

Figure 6.5: Azure Defender dashboard

Let's take a look at what Azure Defender can do for servers, SQL, storage, and container registries.

Azure Defender for servers

Azure Defender is integrated with Azure Monitor to protect your Windows-based machines. While Security Center presents the alerts and remediation suggestions, Azure Defender collects audit records from machines, including Linux machines through auditd, a common Linux auditing framework that lives in the kernel.

Your organization can benefit from using Azure Defender for servers in multiple ways, as it provides threat detection and protection capabilities, such as the following:

- Integration with Microsoft Defender for Endpoint, so when there's a threat detected, it triggers an alert that is shown in Security Center.
- Use the Qualys vulnerability scanner, which we'll discuss in detail shortly.
- JIT access can be used to lock down inbound traffic to your virtual machines, reducing exposure to attacks.
- Change monitoring or file integrity monitoring, which can help you examine files and registries of a given operating system, applications, and changes in file size, access control lists, and the hash of the content that might indicate an attack.
- You can use adaptive application controls for an intelligent and automated solution to create an allow list of applications that can run on your machines, and get security alerts if there are any applications running that are not defined as safe.
- Adaptive network hardening can help you improve your security posture through **network security group (NSG)** hardening.
- Security Center can assess your containers and compare the configurations with the **Center for Internet Security (CIS)** Docker Benchmark.
- You can use **Fileless attack detection** and get detailed security alerts with descriptions and metadata.
- Linux auditd alerts and Log Analytics integration can be used to collect enriched records that can be aggregated into events.

Azure Defender for SQL

Azure Defender for SQL can be utilized to protect your workloads against potential SQL injection attacks, anomalous database access and query patterns, and suspicious activity. It can be used for IaaS- and PaaS-based services. Simply put, it supports the following two scopes:

- **SQL database servers**: This includes Azure SQL Database, Azure SQL Managed Instances, and Dedicated SQL pools in Azure Synapse.
- **SQL servers on machines**: This extends protection for Azure-based SQL servers but also other cloud-based and on-premises environments, such as SQL Server on Azure virtual machines, on-premises SQL Server running on Windows machines, and Azure Arc enabled SQL Server.

Azure Defender for Storage

For remote workers accessing applications in the cloud that have the ability to upload files, we need to ensure that we can determine whether uploaded files are suspicious. Through Azure Defender for Storage, we can use hash reputation analysis and trigger alerts in case there are any suspicious activities or anomalous behavior, or if there is any potential malware being uploaded.

Azure Defender for Storage will display an alert and can email the owner of the storage for approval to delete these files.

Azure Defender for container registries

With Azure Container Registry you can build, store, and manage your container images and artifacts in a managed, private Docker Registry service.

You can scan vulnerabilities in your container images with Azure Defender and review the findings—categorized by severity—in Security Center's list of security recommendations.

Images can be scanned in three different phases: on push, images recently pulled, and on import. These images are pulled from the registry and then run in an isolated sandbox with the Qualys scanner. Then, Security Center will present the findings.

Qualys vulnerability scanners

Azure Defender includes three Qualys scanners: one for machines, one for container registries, and one for SQL. The Qualys scanners, as shown in *Figure 6.6*, are integrated into Security Center and monitor your machines through an extension installed on the actual resource. Then, Qualys' cloud service performs the vulnerability assessment and sends its findings to Security Center:

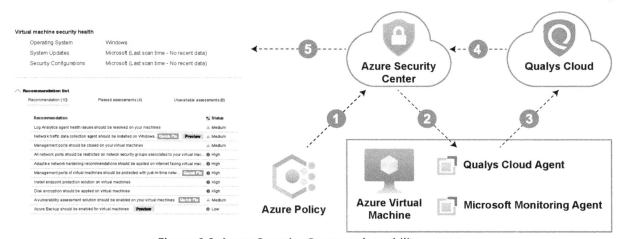

Figure 6.6: Azure Security Center vulnerability scanner

The vulnerability scanner works as follows:

1. Once Azure Defender is enabled for your subscription, you can use the recommendation *vulnerability assessment should be enabled on virtual machines* to deploy the integrated vulnerability scanner extension.

2. Once the extension is installed, the agent collects the necessary security information including the operating system version, open ports, installed software, environment variables, and metadata associated with the files. A scan occurs every 4 hours, and all data is sent to the Qualys cloud service to be analyzed.

3. Qualys analyzes the information and builds the findings per machine. These findings are sent to Security Center.

4. Then you can review the machine's vulnerabilities from Security Center's recommendations page.

It is possible to scan non-Azure machines by connecting them to Security Center with Azure Arc then deploying the extension as above. For large deployments, you can use **Azure Resource Manager (ARM)** remediation scripts, PowerShell, Azure Logic Apps, or the REST API.

Alerts

Since Security Center can be utilized to protect your resources deployed on Azure, other clouds, and on-premises, it can generate multiple types of alerts that are triggered by advanced detections available through Azure Defender.

These security alerts are generated if there are threats detected on any resource. Security Center provides you with a single view of an attack campaign that contains the incidents. A security incident is a collection of related alerts.

You will be able to inspect the security alerts and review more related details, as shown in *Figure 6.7*:

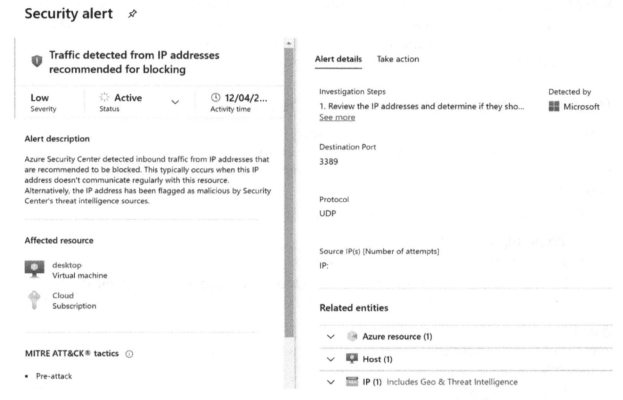

Figure 6.7: Security alert details

As your environment continues to grow, it is more likely that you will have a higher number of alerts shown in Security Center and you will need to have better control of your security alerts and prioritize them accordingly. Security Center assigns a severity to these security alerts in four categories:

- **Informational**: You will see these types of alerts if you drill down into a security incident that is made up of a number of alerts.
- **Low**: These types of alerts might indicate a blocked attack or a false positive with low impact.
- **Medium**: These alerts indicate that a resource might be compromised due to suspicious activity. For example, a sign-in from an anomalous location.
- **High**: A high alert indicates that there's a high probability that your resource is compromised, and you should take action right away.

Security Center allows you to export alerts in different ways. You can download a CSV report on the alert dashboard, configure continuous export, or use an SIEM connector (such as Azure Sentinel) to stream security alerts.

While Azure Defender can help you to protect and remediate potential vulnerabilities, it is also important to implement best practices on the configuration of the resources that are critical for your workloads, such as network resources.

Best practices for securing your network

In the previous section of this chapter, *Security operations excellence*, we reviewed the Zero Trust principle. This assumes that breaches are inevitable and therefore we must ensure we have the right controls in place to verify each request. Identity management plays a critical role in this.

There are three principal objectives for securing your network in the Zero Trust model:

- Preparing to handle attacks in advance
- Reducing the attack surface and the extent of damage
- Strengthening your cloud footprint, including resources and configurations that are part of your environment, to reduce the chances of getting compromised

If your organization is enabling remote work and implementing an end-to-end Zero Trust framework, there are some best practices to adhere to in order to achieve these objectives:

- Network segmentation
- Threat protection
- Encryption

Let's take a closer look at each of them.

Network segmentation

Protecting your corporate network is a top priority, but this is tougher than ever as remote work redefines the security perimeter. It is no longer defined by the physical locations of an organization but now extends to every endpoint that accesses corporate data and services. Therefore, relying on traditional firewalls and VPNs is not enough to protect this new digital estate.

It is known that there's no one-fits-all architecture design for organizations; every single organization has different business needs to be addressed. Taking a Zero Trust approach can help ensure optimal security for your remote workers without compromising the application experience.

Traditional enterprises used to secure their corporate perimeters using traditional network controls. Traditional networks tend to have few network security perimeters and a flat open network, which makes it easy for attackers to move quickly across the entire network. Zero Trust networks are fully distributed with micro-segmentation. Granular network segmentation removes trust from the network, which dramatically minimizes lateral movement and data exfiltration.

Zero Trust networks employ machine learning threat protection and context-based filtering, which can thwart even the most sophisticated attacks. In traditional networks, not all traffic is properly encrypted, which makes them more susceptible to man-in-the-middle attacks, eavesdropping, and session hijacking. In Zero Trust networks, all traffic is encrypted using industry standards to ensure data in transit remains confidential.

The main goal is to move beyond a centralized network to a more comprehensive and distributed segmentation using micro-perimeters. This way, applications can be partitioned across multiple Azure virtual networks and connected through a hub-spoke model. In addition, it is recommended to deploy Azure Firewall in the hub virtual network to inspect all traffic.

To support this approach, Microsoft Azure provides you with multiple cloud-native networking components to enable remote work and mitigate network issues. Organizations are using third-party NVAs available through the Microsoft Azure Marketplace to provide critical connectivity across multicloud and on-premises environments.

Threat protection

Cloud applications that are exposed to the internet are at a higher risk of attacks and therefore we must ensure we scan all the traffic going through them. Threat protection involves the ability to mitigate threats of known and unknown attacks.

Unknown attacks are mainly threats that don't match against any known signature. For known attacks, in most cases, there's a signature available and we can ensure each request is checked against them.

Services available in Azure, such as **Web Application Firewall (WAF)**, can be utilized to protect HTTP(S) traffic. It is possible to use Azure WAF along with Azure Front Door or Azure Application Gateway, while Azure Firewall can be used for threat intelligence-based filtering at Layer 4 of the OSI model.

Encryption

Azure enables you to ensure protection for sensitive data. It provides support in multiple areas of encryption including at rest, in transit, and secrets management. Encryption of data at rest can be applicable to persistent storage resources such as disk or file storage and is available for services across SaaS, PaaS, and IaaS.

Through server-side encryption, you can use service-manager keys or customer-manager keys using Key Vault or customer-controller hardware. Client-side encryption includes data handled by an application running on-premises or outside Azure, and data that is encrypted when received by Azure.

In a Zero Trust approach, we must ensure that all traffic from the user to the application is encrypted. We can accomplish this objective by using Azure Front Door to enforce HTTPS traffic for applications exposed to the internet and utilize networking services and features such as VPN Gateway (P2S or S2S).

Also, if you're enabling access to resources in Azure through virtual machines, you can secure the communication using Azure Bastion, as discussed in *Chapter 5, Enabling secure, remote work with Azure AD and WVD*.

Now that we have reviewed best practices to secure your network and traffic across your environment, it is important to also have a mechanism to get insights from the resources that are part of your environment and be able to proactively analyze potential threats in your environment.

Modernize your security operations with Azure Sentinel, a cloud-native SIEM

Organizations that enable remote work need a way to monitor and get more insights from their environments, whether they are living in the cloud or on-premises.

Azure Sentinel is a cloud-native **security information and event management (SIEM)** service that allows you to simplify data collection from multiple data sources, including on-premises and multicloud environments, through built-in connectors so that you can proactively analyze potential threats and simplify security operations.

Azure Sentinel correlates the security logs and signals from all sources across your applications, services, infrastructure, networks, and users. Azure Sentinel can identify attacks based on your data and places them on a map so you can analyze all the traffic.

First of all, you need to connect your resources, which can include networking components, applications, and data sources. Azure Sentinel provides you with a wide range of connectors to pull data from, including Microsoft services such as Microsoft 365 Defender, Azure AD, Microsoft Defender for Identity (formerly Azure ATP), Microsoft Cloud App Security, and Azure Defender alerts from Security Center, as seen in *Figure 6.8*:

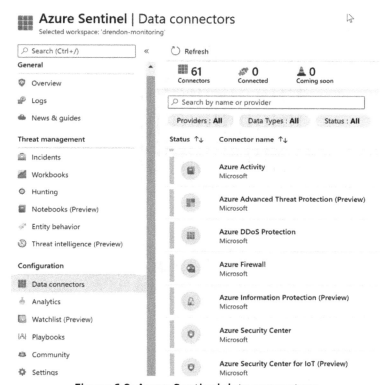

Figure 6.8: Azure Sentinel data connectors

Alternatively, you can choose from third-party solutions or create your own custom connectors, as Azure Sentinel supports Common Event Format, Syslog, and the REST API to connect your data sources. All the data collected is stored in a Log Analytics workspace, which is a container where the data is collected and aggregated. When you enable Azure Sentinel on your subscription, you will be able to select a specific Log Analytics workspace.

Once you connect your resources, you will be able to select a specific workbook that provides a canvas for data analysis and enables you to create custom visual reports within the Azure portal. You can create your own workbooks according to your business needs, and combine data from multiple sources in a single report along with custom visualizations.

Through the **Azure Sentinel** dashboard, as seen in *Figure* 6.9, you can get a high-level overview of the security posture of your organization as it includes events and alerts over time, incidents by status, and potential malicious events:

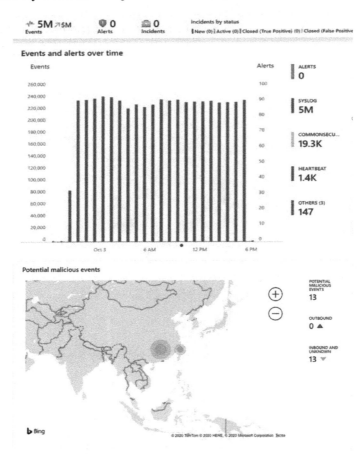

Figure 6.9: Azure Sentinel dashboard

Let's explore some more features that Azure Sentinel provides: threat detection and hunting, investigation and incident response, and managing multiple tenants.

Enable threat detection and hunting

Stream threat indicators your organization is using to Azure Sentinel to enhance your security analysis and hunt for security threats across your organization's data sources, whether they live in Azure, in other clouds, or on-premises.

Azure Sentinel provides you with a hunting capability, shown in *Figure* 6.10, that includes built-in queries that can filter providers or data sources to help your security analysts find issues quickly in the data you already have. Each query provides a detailed description for the use case and you can customize your queries as needed:

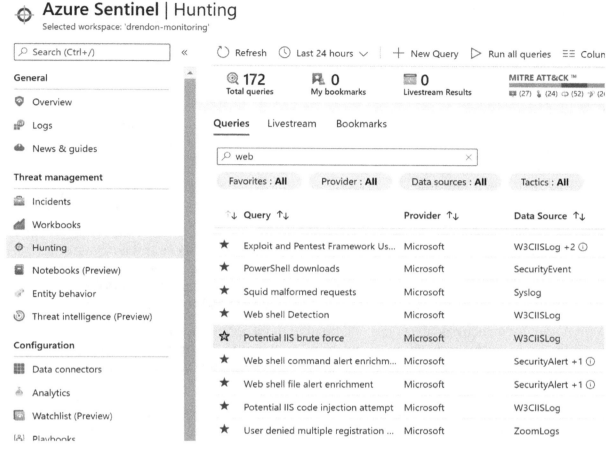

Figure 6.10: Azure Sentinel Hunting

Queries are based on **Kusto Query Language** and fully integrated with Azure Monitor. Once you customize a query that provides high-value insights into potential attacks, you can save it and then use notebooks to run automated hunting campaigns.

Investigation and incident response

Azure Sentinel allows you to see incident operations over time. Incidents are created based on analytic rules and includes relevant evidence of an investigation; therefore, they can contain multiple alerts. If analysts in your organization are looking to investigate incidents, Azure Sentinel provides incident management capabilities that can be accessed through the **Incident** page. You can review how many incidents you have open, in progress, or closed:

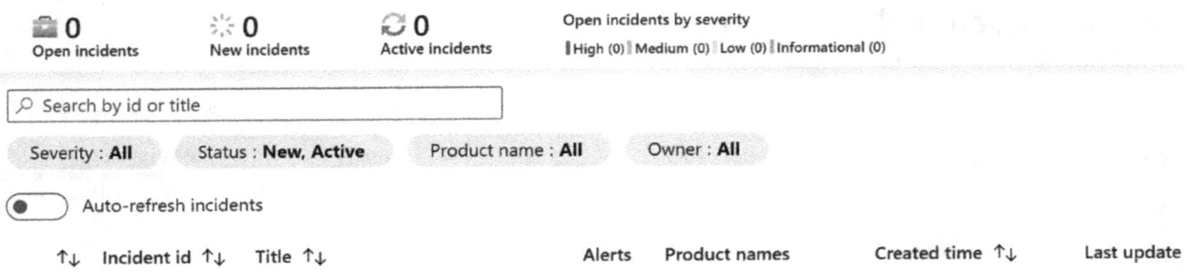

📁 **0**
Open incidents

✳️ **0**
New incidents

🔄 **0**
Active incidents

Open incidents by severity

| High (0) | Medium (0) | Low (0) | Informational (0)

🔍 Search by id or title

Severity : **All**   Status : **New, Active**   Product name : **All**   Owner : **All**

⬤ Auto-refresh incidents

↑↓ Incident id ↑↓   Title ↑↓   Alerts   Product names   Created time ↑↓   Last update

Figure 6.11: Azure Sentinel incident table properties and filters

You can drill down on each incident and see the timestamp, along with the status and severity of the incident. It is also possible to use the investigation graph, which provides an illustrative map of the entities related to the alert and the resources associated with it.

However, the objective is to be able to respond to a security alert and automate responses to these security alerts as much as possible. Azure security playbooks can help you orchestrate and automate your response. These security playbooks are based on Azure Logic Apps and can improve the way you respond to a security alert.

When you get a security alert, you will be able to either run a playbook manually or automate your security playbook. To manually execute a security playbook, you can go through the **Incidents** page and, in the **Alerts** tab, configure which playbook you want to run from the list of available playbooks.

The automated response involves the configuration of a trigger from the security alert. You can configure the action of triggering a security playbook when there is a match on your security alert.

Manage multiple tenants/multi-tenant environments

A huge benefit of using Azure Sentinel to modernize your security operations is the extensibility to manage your customer's Azure Sentinel resources from your own Azure tenant, so there's no need to connect to your customer tenant. This ability for multitenancy is really helpful for organizations that act as managed security service providers to other organizations or customers.

This ability to manage multiple tenants in Azure Sentinel from your own Azure tenant is facilitated by Azure Lighthouse, which enables multi-tenant management capabilities and enhances governance across resources and tenants. If your organization provides managed services to multiple customers, you can use more comprehensive management tooling, delegate resource management, view cross-tenant information from the Azure portal, use ARM templates to onboard customers, perform management tasks, and also provide managed services using the Azure Marketplace.

Based on industry verticals and profiles, organizations must achieve regulatory requirements that dictate recommendations on security controls that have to be adopted such as PCI-DSS, NIST, ISO27001, and so on.

Meeting these regulatory compliance obligations can be a significant challenge in a cloud or hybrid environment. In the final section, we will go through the features available in Azure that can help you enforce these organizational standards.

A unified SecOps experience

As organizations deploy cloud applications and provide access for remote workers, maintaining consistency to ensure cloud compliance, avoiding misconfigurations, reducing the risk of potential attacks, and achieving organization-wide governance can become a huge challenge.

You can enforce organizational standards to assess compliance at scale using Azure Policy and evaluate compliance by evaluating the properties of your resources through policy definitions. Azure Policy provides a compliance dashboard that can be used to see more granular details per resource or per policy, which ensures your resources are compliant. In addition, you can define Azure initiatives, which are a collection of Azure policy definitions, to help you achieve your goals and simplify the management of your policies:

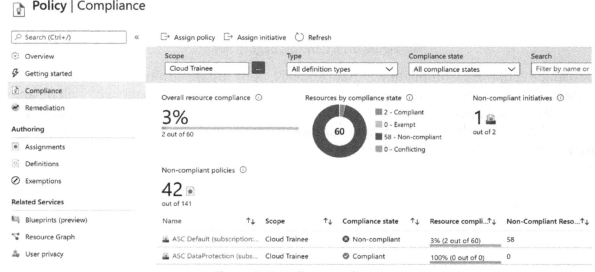

Figure 6.12: Policy compliance

Within this context, your organization can utilize a new capability in Azure Policy called **Regulatory Compliance**, in preview at the time of writing, which provides you with built-in initiative definitions to guide you on the controls and compliance domains. This means you are able to see which compliance domain is intended to be covered, whether by your organization, Microsoft, or a shared compliance domain.

You will see a new tab in the **Compliance** page called **Controls** that allows you to filter by compliance domain and check the details per row:

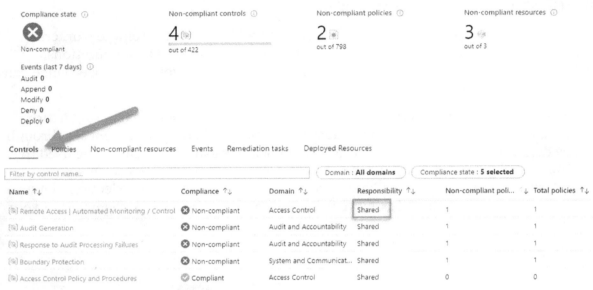

Figure 6.13: Controls page in Policy

By using Regulatory Compliance in Azure Policy, your organization can achieve the compliance required for your business needs. On the compliance page, you will be able to see if a compliance domain is the responsibility of Microsoft, the customer, or if it is a shared responsibility. You can view additional details of audit results if the control you are reviewing is the responsibility of Microsoft.

Summary

Enabling remote work makes security operations more challenging as there's a significant increase in the number of endpoints and environments that need to be monitored and protected. Through the adoption of a Zero Trust strategy, we can reduce the risk of potential attacks and compromises.

In this chapter, we have reviewed some of the core services that Microsoft provides to help you adopt a Zero Trust strategy in your organization and respond to threats quickly and intelligently.

We reviewed how you can empower your security operation teams using Azure Sentinel, a cloud-native SIEM, to help security operations teams stay ahead of their adversaries. We also saw how you can use Azure Security Center to monitor almost all the resources that are part of your environment and protect them without any additional deployment, and use Azure Defender to protect your workloads whether they are in the cloud or on-premises.

In the next chapter, we will see how you can optimize cloud costs by using tools available in your Azure subscription to track resource usage, manage costs across all your environments, and implement governance policies for effective cost management.

Offers, support, resources, and tips to optimize cost in Azure

In the previous chapter, you saw the fundamentals of protecting your applications against cybercrime. We will now focus on various offers provided by Azure and the support you get from Microsoft to optimize costs, various tools that can be used to bring down costs, and tips for managing these costs.

As organizations begin their journey to the cloud, one of the main things they have to learn is how to optimize their costs. From a customer point of view, cost forecasting can be based on how to reduce their datacenter footprint, OPEX pricing models, staff productivity, and other factors.

In this chapter, we will cover the following:

- Understanding and forecasting your costs
- Strategies for optimizing your costs

The first step is understanding and planning for the costs your organization faces.

Understanding and forecasting your costs

Organizations should perform timely analysis on their Azure invoice to monitor costs and reduce the overhead required to manage their IT resources. Using tools such as Azure Cost Management and Billing and Azure Advisor, you can analyze, manage, and optimize your workloads. Before we take a look at these tools and the benefits they provide, we must first understand the factors that are most critical for forecasting your costs in a cloud infrastructure.

Cloud economics

It is important for businesses to understand cloud economics so they can forecast their costs and maximize their cloud investment. Businesses should analyze what their **return on investment** (**ROI**) should be when migrating from an on-premises workload to the cloud or switching from other workloads to Azure. Primarily, you will need to benchmark the cost of running your current datacenter. This can include capital costs, maintenance and operational costs, and software licensing costs. You should also determine the cost of migrating your IT operations to the cloud.

To sum up cloud economics, your organization's focus should be on the following eight-point formula:

1. Reduce your datacenter footprint by streamlining operations.
2. Focus on utilizing on-demand cloud benefits as part of an operational expenditure model.
3. Increase productivity by freeing up your staff from maintenance tasks.
4. Strive for business sustainability.
5. Explore your scalability options and deliver resources when they're needed.
6. Meet security and compliance standards.
7. Use high-availability cloud infrastructure to ensure business continuity.
8. Optimize workload and usage costs.

Azure offers solutions to help you toward all these goals, which we will explore in the following sections.

Cost management tools

Azure has a wide range of tools to provide a great developer and operations experience and to help with understanding the cost of the resources being used in Azure. All the cost management solutions discussed in this chapter have a business impact if utilized properly. Let's take a look at the list of tools available to you in Azure.

Azure Cost Management and Billing

With its full set of cloud cost management capabilities and a single, unified view across all your clouds, Azure Cost Management and Billing helps you increase organizational accountability and manage your cloud spending with confidence. The **Cost Management and Billing** service enables you to check out billing scopes, drill down into your cost management, review your management groups, and diagnose and troubleshoot problems:

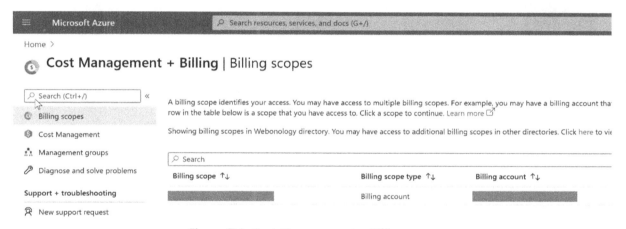

Figure 7.1: Cost Management + Billing page

Management groups are a unique way to group subscriptions and costs. While this chapter is not on management groups, it's worth noting that they're still a useful Azure resource to manage your subscriptions.

Organizations need to take a unique approach to provide access to cost management to users that aren't in the financial part of the organization. There are ways to allow these cost structures to be visible to your organization's operations and development side. Power BI dashboards and APIs can be used to pull reports and create neat, unique dashboards to access cost information; you can read more about this here: https://docs.microsoft.com/power-bi/connect-data/desktop-connect-azure-cost-management. As well as Power BI, Azure allows integration with custom-written apps, as well as SaaS applications like CRM for digesting and using your data.

As mentioned, management groups can help provide a better view of costs, and you can use tags to assign the budget codes within an organization. The following screenshot demonstrates how costs can be visualized on a service basis:

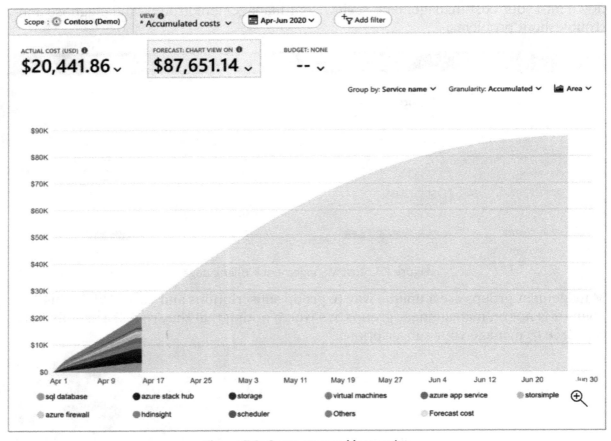

Figure 7.2: Costs grouped by service

With Azure, your organization can easily manage the costs of hybrid cloud environments (Azure and AWS, say) from a single location, getting the best insights from data from both clouds.

Let's take a quick look at the **Cost Management** section for the pay-as-you-go model (a pricing model we will discuss later in the chapter):

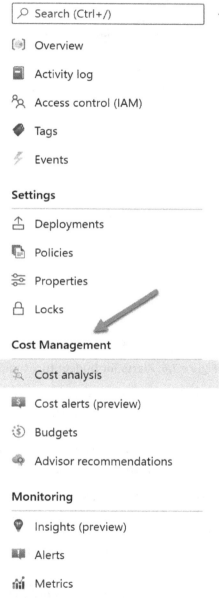

Figure 7.3: Cost Management section

As you can see, the **Cost Management** section allows you to analyze costs, create cost alerts, assign budgets, and get advisor recommendations. When you click on **Cost analysis**, you'll get a breakdown of the resources within the resource group's scope, such as data stores or serverless accounts.

You can also set cost alerts, which is an excellent tool to help monitor overspend on non-production environments, by setting a threshold and then receiving alerts. Alerting can be used in conjunction with budgets, as shown in the following screenshot:

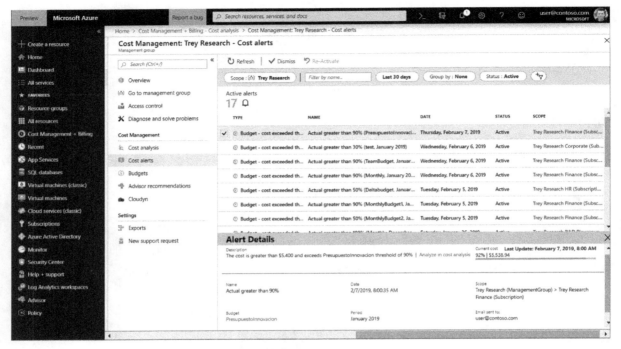

Figure 7.4: Alerts for budget threshold

Budgets and alerts are an invaluable next step for organizations after they have visualized their current spending. You should monitor and analyze your Azure bill to ensure there aren't any charges or hidden infrastructure that have made their way through. This is generally essential for non-production environments, where things are a little bit more loosely controlled than in production.

When an alert is triggered, you should undertake a cost review to isolate the causes and determine whether any action needs to be taken. This could be in the form of revising your budgets or implementing additional Azure Policy controls, to name a few. Assigning budgets also helps with controlling the costs of your different teams' projects and business units.

To read about Azure Policy and its important role in governance, visit https://docs. microsoft.com/azure/governance/policy/overview.

To find out more about the capabilities of Azure Cost Management and Billing, visit https://docs.microsoft.com/azure/cost-management-billing/cost-management-billing-overview.

Azure pricing calculator

The Azure pricing calculator is a free tool that can be used to get real-time cost estimates of subscription services. You can personalize your view of the estimates through a central dashboard. The pricing calculator can be combined with other utilities, such as the Azure **Total Cost of Ownership (TCO)** calculator, for better cost optimization. We can create estimates and save them to our accounts, or share them with relevant parties.

Azure regions can sometimes affect the cost of resources in Azure, so it is helpful to use the Azure pricing calculator to predict costs before selecting a resource:

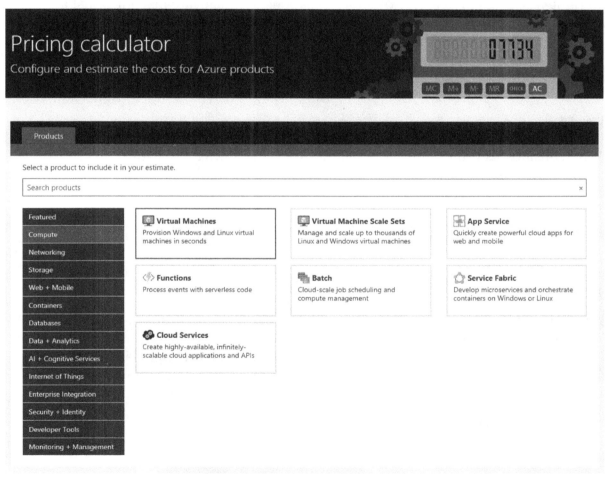

Figure 7.5: The Azure pricing calculator

The Azure pricing calculator can be found here: https://azure.microsoft.com/pricing/calculator/.

One of the hurdles organizations can run into when using the pricing calculator is factoring in the infrastructure around the resources, they are looking for to get the right estimate. You should always take into account all of the network infrastructure needed to deploy your applications to the cloud.

For example, consider an organization with two VMs, one running a website and the other running services to support that website, which wants to move both VMs out to the cloud. They go to the pricing calculator, input the two VMs in the pricing calculator, and discover their average monthly cost.

One of the things that they should have done in this scenario is look at the VMs' usage to right-size them in Azure. A lot of organizations over-allocate machines that are needed for future expansion, but in the cloud, you can expand any time you need to, so you don't need to put oversized resources into play. The organization also missed the routers and firewalls and didn't take into account VM scale sets, or the ability to protect their infrastructure should a VM go down. As you can see, a good plan can go a long way.

Total Cost of Ownership calculator

When businesses want to take a bigger, holistic view of what it means to migrate their Azure solutions, they can use the TCO calculator. This provides detailed information about moving on-premises workloads to the cloud and what it will take to make the transition smoother.

You can find more information about this calculator at https://azure.microsoft.com/pricing/tco/calculator/.

Azure Migrate

Another tool to use if you're choosing to migrate from on-premises to the cloud is Azure Migrate. Azure Migrate provides a centralized single pane of glass view to assess and migrate on-premises servers, infrastructures, applications, and data to the cloud. This assists organizations by providing a wide range of tools and options to help the process of assessing and migrating these different resources and workloads to Azure.

For more information on how to use Azure Migrate, go to https://docs.microsoft.com/azure/migrate/migrate-services-overview.

Choosing the right pricing model

As well as tools, it is equally important to understand the various pricing models Azure offers to better forecast your resources and their costs. In Azure, there are two standard pricing models for services:

- The pay-as-you-go model, also known as the consumption-based model
- The fixed-price model

Most Azure resources, such as PaaS or serverless, will use the consumption-based model. An example of fixed pricing is Azure **Reserved Virtual Machine Instances (RIs)**, or **Azure Reservations** more broadly.

The fixed-price model is more predictable, while the pay-as-you-go model can be variable, so you need to consider business requirements when choosing between them. We'll go into detail about both models in the following sections.

Consumption-based model

The consumption-based, or pay-as-you-go, model follows a utility-based structure. This means that you pay for what you use and there is no monthly or fixed cost for the resources. An example of this would be a function application, where you'd pay one dollar for a million uses of the app.

This pay-for-what-you-use model can help you control costs, but could also potentially increase them. An example of this might be something like CosmosDB, which is based on usage or request units and can become very costly at the end of the day depending on how many times queries are run.

You can use architectural patterns like base load-leveling and autoscaling of services to help with fixed minimum performance levels. If you need burst workloads, you can use something like a throttle pattern to maintain the quality of the service under load.

Fixed-price model

Under a fixed-price model, you pay for a resource whether it is used or not. We already mentioned Azure reservations and specifically RIs, but another example would be an App Service plan in Azure. When you select a type of resource you incur a monthly cost, and regardless of how many people utilize the application, you still pay this cost. This costing model helps organizations predict their spending in Azure a little better, but it can also be a problem based on utilization.

Organizations buy reserved instances when they plan to make extensive use of VMs for their applications. In return for committing to resources for fixed terms, Azure can provide them at discounted rates. When you first migrate to the cloud, there may be uncertainty around your resource usage, so depending on your workloads, it may be best to get a handle on this before you consider RIs.

There is a trade-off between scalability and predictability when keeping costs structured, and most organizations opt for a blended approach of the two pricing models.

For more information on pricing models, head over to https://docs.microsoft.com/azure/architecture/framework/cost/design-price.

Now that we have an understanding of the various ways Azure can help you to assess and keep on top of your costs, we'll move on to optimizing these costs.

Optimizing your costs

Cost optimization is the process of reducing your resource costs by identifying waste, mismanaged resources, and right-sized resources, and also by reserving capacity. With cost management being one of the five principles of cloud governance, it is important to focus on ways of establishing a spending plan and budget for your cloud resources. You will also want to use monitoring and alerting to enforce these budgets and detect anomalies that emerge from development practices. Microsoft provides the Cloud Adoption Framework for guidance on this, which details tools, best practices, and documentation to help organizations succeed in the cloud. You can find more information at https://docs.microsoft.com/azure/cloud-adoption-framework/.

A general method for estimating cost and making changes to the pricing models we discussed earlier is by monitoring your workloads on a peak throughput. For example, if the utilization is high all the time, a pay-as-you-go model would be less efficient for baseline cost estimation. Hence, while providing flexibility, the pay-as-you-go model alone doesn't always constitute a cost-saving.

As has already been pointed out, the best cost optimization is having a plan, so you should use the tools at hand, from pricing calculators to checklists, to establish your policies, budgets, and controls to limit the spending on your solutions. The Azure Well-Architected Framework is a great starting point for building these policies and a good source of information. It is located at https://docs.microsoft.com/azure/architecture/framework/cost/. We discussed the Azure Well-Architected Framework in *Chapter 4, Cloud migration: Planning, implementation, and best practices*.

As we begin exploring the optimization options Azure provides in the next few sections, you should keep in mind that it's not just about moving a workload to Azure, but all the things surrounding it as well. When migrating to the cloud, many organizations miss some of the associated underlying infrastructure costs.

Azure Advisor

One of the useful services you can access through Azure Cost Management is Azure Advisor. Azure Advisor analyzes your resource configuration and usage, providing personalized offers and recommendations so you can optimize your resources for cost-effectiveness, security, and performance. For example, it helps to point out unused resources or those that have been idle for a long time, including right-sizing resources like SQL.

For more information on Azure Advisor, visit https://docs.microsoft.com/azure/advisor/advisor-overview.

It is important to point out that you should right-size your resources for the cloud. Most organizations overbuy internal infrastructure only for the purposes of future growth. When moving to Azure, you can scale horizontally as well as vertically, so if your machine needs more resources, you can add more. You can also use a hybrid approach, where you don't need to move all of your workloads out to the cloud—you can move the pieces of the workload that need more performance or more scalability at the instance level.

Azure Hybrid Benefit

Azure Hybrid Benefit is a licensing benefit that drastically lowers the cost of running workloads in the cloud. It will allow you to use on-premises software assurance-enabled Windows or SQL Server licenses, as well as Red Hat and SUSE Linux subscriptions, on Azure. These types of benefits can drive down costs by 40% or more. Furthermore, you are given three additional years of security updates for free when you move your Windows Server or SQL Server 2008 and 2008 R2 workloads to Azure.

One of Microsoft's primary goals of the Hybrid Benefit model is the savings that come from using a Windows or SQL Server Hybrid license, which drives down the cost of a managed instance by almost 85%. These benefits can apply to VM operating systems as well. *Figure 7.6* shows some potential cost savings involving RIs and Azure Hybrid Benefit:

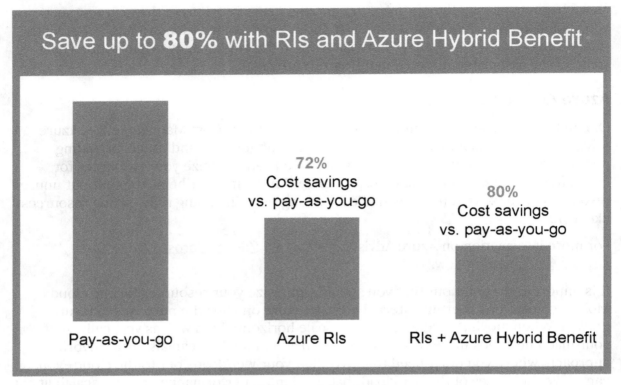

Figure 7.6: Potential cost savings

It is recommended that you use the Hybrid Benefit calculator to figure out your overall cost savings using the hybrid model, which can be found at https://azure.microsoft.com//pricing/hybrid-benefit/#calculator.

Azure reservations

Azure reservations can help you save money when you commit to a one- or three-year cost for your resources. These commitments can save up to 72% of costs over the pay-as-you-go pricing, since this is a contract in time to use these resources that you pay for upfront. Reservations should be used when you have long-term application plans in play that fit the timescales. They are only applied to specific resources in Azure, such as VMs.

To learn more about reservations, visit the following links:

- https://azure.microsoft.com/reservations/
- https://docs.microsoft.com/azure/cost-management-billing/reservations/save-compute-costs-reservations
- https://azure.microsoft.com/pricing/reserved-vm-instances/

Spot Virtual Machines

Spot Virtual Machines allows you to use unused Azure compute capacity at deep discounts. The caveat is that when Azure needs the capacity back, it will evict your Spot VM. For this reason, Spot VMs are well-suited for workloads that can be interrupted, such as batch processing, non-production environments, big data analytics, and container-based and large-scale stateless applications. The available capacity varies from region to region, depending on the time of day or size of the workload; as a result, these VMs have no SLA. You can view the price history and the eviction rate for the Spot VMs you select in the Azure portal.

If you would like to learn more about using Spot VMs, visit https://docs.microsoft.com/azure/virtual-machines/spot-vms.

DevTest pricing

Azure DevTest enables developers to create VMs or other PaaS resources without needing fast approval. It allows teams to develop non-production environments to test applications rapidly and enables organizations to put a consistent budget on these non-production resources or sandbox areas. In addition, it gives you the ability to set auto-startup and shutdown schedules for VMs in order to cut costs and allows policies to be put in place to ensure larger resources don't get used.

To learn more about how to use Azure DevTest, go to https://docs.microsoft.com/azure/devtest-labs/devtest-lab-overview.

Optimization checklist

Based on what we've learned in this chapter, here are some ways to optimize your costs in Azure that you can implement today:

1. Right-size your unused resources because they can scale both horizontally and vertically in the cloud.

2. Shut down your unused resources when they aren't needed. This can be handled through automation scripting.

3. Take advantage of hybrid models in the cloud, only moving the necessary workloads to the cloud. Get more information at https://docs.microsoft.com/azure/cost-management-billing/costs/tutorial-acm-opt-recommendations.

4. Use reserve instances for the large VM resources that you know you're going to use for an extended period of time in the cloud.

5. Set up budgets, allocate costs to different groups, and get alerts when you get close to these budgets.

6. Set up Azure DevTest environments for development and testing before publishing to your public workloads.

7. Explore other Azure services like PaaS, SaaS, or serverless to optimize your approach to the cloud and control costs.

8. Cost-optimize your workloads, following your Azure Advisor best practice recommendations.

9. Review the architecture of your workloads using the Microsoft Azure Well-Architected Review assessment and design documentation to see where you can optimize costs.

10. Take advantage of Azure offers and licensing terms such as Azure Hybrid Benefit, paying in advance for predictable workloads with reservations, Azure Spot Virtual Machines, and Azure Dev/Test pricing. For more information on Azure offers, visit https://azure.microsoft.com/support/legal/offer-details/.

Summary

As you have seen, it is beneficial for organizations to put organization-wide cost management and guardrails in place to help free up their teams so they can provide more innovation. Balancing your workloads for performance and resiliency while trying to maintain costs in the cloud can be cumbersome, and you shouldn't allow costs to be the only driving force.

In this chapter, we have reviewed ways to optimize your costs, from right-sizing or creating automation around unused resources to building budgets and alerts for your teams to monitor expenses.

With Azure, there is a distinct cost difference between on-premises datacenters and virtualizing those datacenters in the cloud. We reviewed some of the ways to approach these costs and cost models in Azure. We talked about the unique benefits of adopting Azure, including the ability to span to the cloud with hybrid models for workloads that don't entirely need to run in the cloud. We have also seen some tools to help manage and optimize costs.

In summary, remember to always:

- Choose the Azure compute services appropriate to your needs
- Right-size resources
- Shut down resources that are not being used
- Configure autoscaling
- Reserve instances
- Use Azure Hybrid Benefit
- Set up budgets for different teams and projects

We shouldn't allow costs to drive our use of Azure but be aware of those things that can affect the bottom line.

8
Conclusion

Thank you for reading the *Azure Strategy and Implementation Guide, Fourth Edition*. Throughout the different chapters, we have reviewed how you can start your journey in the Azure cloud, use the Microsoft Azure Well-Architected Framework, and adopt best practices to improve the quality of your workloads in the cloud. There are many different scenarios possible for running your workloads on Azure to meet your organization's solution needs. We learned how important it is to use design principles and realized how crucial *planning* is when moving resources to Azure.

We hope this end-to-end guide has given you a better understanding of the latest Azure technologies and innovations, how they can help your business, and the framework for your strategy for adopting Azure.

Resources

Here are some helpful resources to help you take the next steps in your Azure migration journey:

- Take a free tutorial on Azure Fundamentals with hands-on exercises: https://docs.microsoft.com/learn/paths/az-900-describe-cloud-concepts/
- Sign-up for an Azure free account to explore infrastructure services from Azure: https://azure.microsoft.com/free/virtual-machines/
- Explore the Cloud Adoption Framework for more insights on how to move to the cloud: https://docs.microsoft.com/azure/cloud-adoption-framework/

Glossary

- **Annualized failure rate (AFR)**: The estimated probability that a device or component will fail during a full year of use.
- **App modernization**: The modernization of an existing IT asset involving either its refactoring or re-architecture, or both. The goals of app modernization are often to produce cost and operational efficiencies in the cloud.
- **Availability Zone**: A fault-isolated area within an Azure region with redundant power, cooling, and networking components.
- **Azure AD Multi-Factor Authentication (MFA)**: A type of authentication in which users are asked to verify their identity through an additional form of identification, such as a fingerprint or a code on their mobile phone.
- **Azure Arc**: Enables you to view and manage compute resources, whether they are on-premises, across multiple vendor clouds, or distributed on the network edge, through the Azure management interface and operating model.
- **Azure Connected Machine agent**: A software package that enables you to manage Linux and Windows machines regardless of whether they are hosted outside of Azure, across multicloud environments, or on-premises.
- **Azure Disk Storage**: High-performance, highly durable block storage designed to be used with Azure Virtual Machines.
- **Azure DevOps**: Azure is all about automation and scaling. You can leverage Azure DevOps to deploy your code and infrastructure in Azure to maintain a stable and consistent deployment process.
- **Azure governance**: Azure governance is a set of guardrails that help organizations with Azure compliance and security policies. Azure governance helps ensure that all parties have their goals aligned and understand their responsibilities in their journey to Azure.

- **Azure Infrastructure as a Service (IaaS)**: A set of computing, storage, and application capabilities provided as a service by Microsoft to support your workloads in the cloud. Azure IaaS offers security and the ability to instantly scale your infrastructure to manage and operate your workloads from anywhere while reducing costs.

- **Azure pricing calculator**: The tool you use to estimate your infrastructure costs before you build.

- **Azure Private Link**: An Azure service that provides private connectivity from a virtual network to Azure **Platform as a Service (PaaS)**, customer-owned or Microsoft partner services.

- **Azure Resource Manager (ARM)**: Resources are configured in Azure through a portal but can also be done programmatically. ARM templates, when integrated with Azure Pipelines, can help you achieve **continuous integration and continuous deployment (CI/CD)**.

- **Azure Security Center**: A tool to improve your security posture. It can protect your workloads whether they live on Azure, on-premises, or in other clouds.

- **Azure Sentinel**: A cloud-native SIEM service with built-in AI for analytics that removes the cost and complexity of achieving a centralized and near real-time view of active threats in your environment.

- **Azure Stack HCI**: An Azure service that ensures a consistent environment for your Linux and Windows workloads through software-defined infrastructure. This hyperconverged infrastructure is ideal for hybrid, on-premises environments.

- **Cost analysis**: Cost analysis is a helpful tool to keep your infrastructure spend under control. You can leverage monitoring and alerting to help you be more proactive and cost efficient.

- **Cybersecurity**: A term commonly used to refer to the measures taken in anticipation to oppose an attack on datacenter infrastructure.

- **Edge computing**: It is not always about remaining connected to the internet. Edge computing allows you to build solutions without the requirement for full-time connectivity.

- **Encryption**: A method of encoding information. The major encryption concerns are handled by Azure, including encryption both at rest and in transit, as well as key management.

- **ExpressRoute**: A service available in Azure to create a private connection between Azure datacenters and infrastructure, usually in a co-location environment.

- **Fixed minimum performance**: In Azure, to right-size your infrastructure, you want to ensure that you have a set minimum performance in mind. This leads to an understanding of the minimum number of resources required to run your application, which will help define your scaling.

- **High availability**: The characteristic of a system or network to ensure operational continuity over a particular period of time while avoiding downtime.

- **Hybrid cloud**: Not everything is a good candidate for the cloud. Hybrid solutions are a way to build applications without fully moving your workloads to the cloud.

- **Hybrid license**: Hybrid licensing is when you leverage an on-premises or pre-purchased license in Azure to help manage costs. This is useful for organizations that have special pricing for on-premises resources.

- **Infrastructure as Code (IaC)**: IaC is what an ARM template generates for infrastructure deployment consistency.

- **Multicloud**: One size doesn't fit all with cloud providers. A multicloud approach enables organizations to deploy their cloud assets, applications, and resources across different cloud providers. Multicloud models are used in order to maintain high levels of uptime for mission-critical applications.

- **Multicloud strategy**: Whether for government or corporate board compliance, multicloud strategies allow an organization to address a number of issues. These can range from avoiding price-lock from a cloud vendor to designing a resilient disaster recovery plan and adeptly managing government regulations of data storage. Whatever the need or want, these strategies enable organizations to use multiple cloud providers.

- **PaaS (Platform as a Service)**: Encompasses the benefits of **Infrastructure as a Service (IaaS)** and can also include middleware, database management, container orchestrators, and **business intelligence** (**BI**) services. PaaS solutions often include pre-coded application components such as security features, directory services, and workflow features.

- **Passwordless**: Any method used to verify a user's identity without requiring the user to provide a password, such as the use of biometric gestures like fingerprints or device-specific PINs.

- **Private cloud**: Cloud computing resources that are utilized by a single organization and not publicly available. The organization hosting the private cloud is responsible for managing and maintaining the infrastructure the cloud resource model is running on.

- **Resiliency**: A network or system's ability to recover from failures while maintaining adequate levels of operability in the face of faults, threats, and challenges.

- **Serverless**: A cloud computing execution model in which compute resources are allocated on demand by their customers' cloud provider. Serverless application environments, serverless functions, and serverless Kubernetes are examples of serverless computing resources available in Azure.

- **Service level agreement (SLA)**: Defines the level of service you expect from a vendor. In this case, the SLAs describe Microsoft's commitments for uptime and connectivity.

- **Shared responsibility**: When moving to the cloud, you have to learn a new shared responsibility for your infrastructure in code. It is good to understand and know where your responsibilities stop and the cloud provider's responsibilities begin.

- **Site Recovery Mobility service**: The Mobility service agent captures data writes on the machine and then forwards them to the Site Recovery process server.

- **Azure Spot Virtual Machines (Spot VMs)**: Spot VMs is an Azure service that lets you buy unused Azure compute capacity (VMs) for your interruptible workloads at large discounts compared to pay-as-you-go prices. It is important to note that Azure can evict Spot VMs when the underlying capacity is needed.

- **Tenant**: A term usually used to represent an organization in Azure AD.

- **Transparent data encryption (TDE)**: TDE is how data repositories are encrypted at rest by default in Azure.

- **Windows Virtual Desktop**: A system for desktop and app virtualization that runs on Azure.

- **Zero trust principles**: Comprises the security model that enables users to work more securely. This includes authentication and authorization, controls to limit access to resources, and breach prediction. Simply put, *never trust, always verify*.

Index

About

All major keywords used in this book are captured alphabetically in this section. Each one is accompanied by the page number of where they appear.

P

PaaS: 2-3, 7-8, 52, 54-55, 57, 72, 88-89, 133, 157, 172, 189, 193-194, 199-200
package: 61-62, 198
parallel: 37, 97
parameter: 14, 18-19, 133
password: 20, 22-23, 109-110, 118, 121-122, 200
patches: 40, 42
path: 34, 67, 77, 141
patterns: 19, 26, 35, 73, 166, 189
PCI-DSS: 132, 177
performance: 22, 34, 38, 41, 51-52, 64, 68-70, 76-77, 82-84, 89, 96, 98-100, 106, 149, 189, 191, 194, 200
perimeter: 135, 153-154, 170
permissions: 23, 92, 125, 159
petabytes: 85
phishing: 121
pipelines: 12-13, 37, 199
plan: 2, 7, 12, 65-66, 73, 79-80, 93, 98, 103-104, 106, 109, 146, 188-190, 200
platform: 1-2, 4, 9, 17, 21, 33, 35, 42-43, 46, 50, 52, 54-55, 67, 85, 88, 96-97, 106, 125, 133, 150, 157-159, 199-200

policies: 23-26, 37-38, 44, 85, 100, 102-103, 111, 122, 125-126, 135, 138, 143, 157, 159-160, 163, 177, 179, 190, 193, 198
pools: 2, 140-141, 166
portal: 8, 11, 13, 20, 22, 35-37, 39-41, 43-44, 54, 56-57, 59-60, 64-65, 67, 90, 99, 111, 113-114, 117, 137, 140-141, 148, 153, 156, 164, 173, 176, 193, 199
portfolio: 31, 72
Postgres: 43
PostgreSQL: 35-36, 39, 42, 48, 73, 101
posture: 9, 22, 101, 106, 151, 153, 155-159, 161-163, 166, 174, 199
PowerShell: 8, 13-14, 36, 54, 57-59, 63, 68, 70, 140, 148, 168
predict: 187, 189
price-lock: 200
pricing: 67, 78, 96, 163, 181, 185, 187-190, 192-194, 199-200
primary: 103, 110, 191
principles: 76-77, 152, 190, 197, 201
private: 4-7, 35, 45, 49, 52, 59, 83, 88, 130, 136-137, 152, 167, 199-200
privileged: 23, 122-123, 162
processes: 12, 24, 35, 38, 50, 53, 55, 71, 103, 154, 159
protocol: 70, 119

provisioning: 2, 40-41, 43, 97, 127
proxy: 123-124, 152
public: 4-7, 35-36, 39, 41, 43, 55, 88, 90, 100, 119, 128, 136-138, 194
Python: 85

Q

Qualys: 166-168
queries: 27, 40, 149, 174-175, 189
queue: 44, 55
quickstart: 54, 119, 150

R

RA-GRS: 101
range: 2, 6, 14, 21, 35, 41, 53, 78, 83, 100, 130, 146, 173, 183, 188, 200
ransomware: 65, 101
RBAC: 26-27, 37-38, 40, 43, 58, 84-85, 140
RDMA: 66, 95-96
real-time: 34, 40, 51, 77, 187, 199
recovery: 6, 63, 65, 70, 76, 100, 103-105, 155, 162, 200-201
redeploy: 8, 56, 60
redundant: 39, 101, 198
region: 39, 43-44, 78, 104-105, 126, 137, 143-144, 193, 198
registry: 38, 40, 44-45, 91, 167